AF699969

Years Touched with Memories

Ward Ritchie

YEARS TOUCHED WITH MEMORIES

FOREWORD BY
LAWRENCE CLARK POWELL

AB BOOKMAN PUBLICATIONS, INC.
CLIFTON, NEW JERSEY

I wish to express my appreciation to Gloria Stuart and Lawrence Clark Powell for their suggestions while reading the manuscript and proofs. Also to Dana Cordrey and his staff at Premier Printing Corporation for their counseling during the production of this book.

On the cover is shown
the pressmark of the Ward Ritchie Press
engraved by Paul Landacre.

Published by
AB Bookman Publications, Inc.
P.O. Box AB
Clifton, New Jersey 07015

Copyright 1992 by Ward Ritchie

All rights reserved. No part of this book may be reproduced in any form by any electronic or mechanical means (including photocopying, recording, or information storage and retrieval) without permission in writing from the copyright holder.

ISBN: 0-9634318-0-3

Designed by Ward Ritchie
Printed in the United States of America
by Premier Printing Corporation, Brea, CA

A Good Long Friendship

IT HAS INDEED BEEN a good, long friendship, nearly eighty years now and still going strong. How did it come to last that long? Few friendships do. Although we've never talked about it, I believe we would agree that books have been the strongest bonding agent – the bookish lives we've shared in nearly every aspect. Printing is beyond me as spelling is beyond him. Who ever had it all? Another good adhesive was an acceptance of our differences of temperament, the one passive, the other positive. That we never fell out was due to him, not me, although once I did get cross with him when he left printing for advertising. What I failed at the time to understand was that wartime economics were the cause of that temporary defection, nor did I appreciate that his skills as a designer-printer made him widely useful in advertising layout and production. The cruelest blow he suffered from the war was the death of his partner, Gregg Anderson, above the Normandy beach, yet that loss of Gregg led him to even greater achievements as a designer. Although Fate moved us around on the map, we always came back together near where we started, now with only 500 miles between seacoast and desert. The chief beneficiary has been the telephone company. Whose nickel it is favors the Scot.

When I asked him what I should emphasize in what could be the last of the preambles written for his works, he suggested those bibulous times shared on the Côte d'Or with Alfred and Mary Frances Fisher. In the weeks before he saw me off from Gare de Lyon on the noon *rapide* to Dijon, we went our separate ways around Paris.

While he haunted the bookshops and printers' studios, I sat on a rented iron chair in the Luxembourg Gardens and began the Great American Novel, no less. Someone had to write it, why not I? We gathered at day's end to eat at Brasserie Lipp, with a cautious glance at a visiting Hemingway and his newspaper cronies nearby. After a tour of the Deux Magots and the Flore, and roasted chestnuts with ruby port from the crone at the corner, we turned in at the Hotel Crystal, where we shared a two dollar room and read aloud from the work in progress. He had more nocturnal staying power and still has, while I was the early riser and still am.

After I settled in at the university, he came for weekend visits. It was only two hundred miles from Paris and halfway to Switzerland. I met him at the station and we made our way through the maze of curving streets to the Pension Rigoulot where the Fishers and I had rooms and board. The family welcomed him with a sumptuous meal, including several bottles brought up from the cellar. Later in the Fishers' top-floor rooms we gave that minister's son his first lesson in crap shooting, which relieved him of any loose francs. Music came from a little wind-up phonograph, my first purchase upon arriving in Dijon. Records included Gershwin's *American in Paris,* Brahms' *First Symphony,* and ten-inch Polydor 78s of *Red Nichols and His Five Pennies.* Their lean and hungry jazz was of the kind my erstwhile Varsity Vagabonds had sought to emulate. The novel had been sidetracked for stories of adventures as a musician on the *President Harrison* and the *Yale.* That was Fisher's advice when he told me that my writing had only the sound of echo. I was still smart enough to learn. Like all uncertain beginning authors, we needed listeners. While Ward read those lyrics that came to form *XV Poems for the Heath Broom,* the more self-assured Fisher lured us toward sleep with excerpts from his epic. As for M.F., she smiled, knitted, and said nought. Her time had yet to come.

Those were the golden years of our *belle jeunesse,* which came fortunately between wars. We did indeed call on the Daughters of Joy, where I played the piano for everyone to dance, until Madame sent the girls back to work upstairs. It was a licensed brothel, clean in every way and owned by the wife of a town councillor. It was more than meals that Madame offered me. What *is* the French word for perks?

If, as he says, I enlarged his social life, so did he expand my mind. Because of him my life went the way it did. Why is it that his pages leave me sad? Perhaps from realizing that all love is prelude to loss. Would that we could go on forever, free of regret, and above all, grateful.

LAWRENCE CLARK POWELL

Tucson
Bajada of the
Santa Catalinas
15 June 1992

Years Touched with Memories

ALTHOUGH I HARDLY REMEMBER THE EVENT, I HAVE BEEN TOLD I WAS BORN IN LOS ANGELES in a small house on Fedora Street near Pico Boulevard. The year was 1905, so I have now lived in each of the decades of the twentieth century. Before I was three years old I was moved to the small town of South Pasadena. My father was a compulsive mover and at this time was seduced by the alluring description of the Oneonta Tract being carved from the orange groves in the southeastern part of South Pasadena. The real estate brochure that influenced him read, "Under the shadows of the wide-spreading oaks soon there will be the low East-Indian bungalow; on some elevated knoll, the Swiss chalet; where oleanders and magnolia bloom, quaint adobes will rear their walls. Gardens shall bloom here, rioting with roses and passion flowers; beauty shall prevail — and hundreds of homeseekers will be eager to seize upon this opportunity of finding such a resting place from the turmoil and stress of business life."

4

Father was so beguiled by the prospect of life in this new paradise that he immediately sold our home in Los Angeles and had us camping out on the grounds as our new house was being built in the midst of an orange grove. It was truly an idyllic place in which to spend one's early years. Our playground was a vast territory. To the east were the fields of San Marino, composed mostly of a few large estates belonging to the Huntingtons, the Pattons, and the Valentines. The Old Mill, or *El Molino Viejo,* was on Huntington land. In earlier days it had been used to grind the grain of the San Gabriel Mission several miles distant. It had gone through many roles since it was built in 1816. Its function as a mill had ceased when its stream had silted up to a mere trickle. During my boyhood it was semi-abandoned, and we found it to be a perfect place to play our Indian games. We were then absorbed by the books of Ernest Thompson Seton and James Willard Schultz and trying to live the life of Seton's "little savages." The Old Mill was a fine rendezvous for us with its thick adobe walls and huge hand-hewn rafters wrapped at the joints with lengthy strips of rawhide. We'd climb about these rafters, occasionally cutting off a piece of rawhide for a slingshot. Nearby was Wilson's Lake, once Vineyard's Lake as part of the large vineyard of Don Benito Wilson. It had originally been fed by the Old Mill Stream, but now that it was but a mere trickle, the lake was not much more than an overgrown marsh. But we had made a raft and could paddle about it.

Closer to home were the endless orange groves that gave us both nourishment as we played and ammunition for our frequent war games. A bit farther afield was Chapman's Ranch, which belonged to the grandfather of Dick Chapman, one of our close pals. We could bicycle out there to swim in the reservoir and, during watermelon time, feast ourselves on ripe melons that we'd pick from the fields. It was a wonderful life, with friends to remember who went on to become lawyers, doctors, and teachers. A few became bookmen, among them Bill Jackson, Larry Powell, and myself.

My first memory of Bill Jackson was when we were in the third grade at the Marengo Avenue School. He had just moved from Canada to South Pasadena, where his father had been given the pastorate of the local Baptist Church. Soon after his arrival, I spotted Bill one afternoon sitting on the school steps with a stack of books. I sat down beside him and began looking through them, curious about where he got them. He told me about the library, where anyone could borrow books. While South Pasadena was little more than a village, the center of town was a couple of miles from where I lived, and I doubt if I had been there more than once in my life. It was unknown territory. Bill lived in the northern part of town and he'd early found the library, which was within walking distance of his home. The next Saturday, on my bicycle, I braved the unknown and pedaled up the long hill until I found the little library that Andrew Carnegie had given our town. It was an unbelievable experience for a nine-year-old boy to see so many books. It wasn't long before the word got around our class and a number of us, including Bill Jackson, Larry Powell, Pat Kelley, Buddy Weldon, and

6

Ward Ritchie at three

I, were competing to see who would be the first to read all of the books in the library. None of us succeeded, but we made a good try.

By the time we were in high school, Henry Huntington, whose estate was just east of town, began building a stately library building to house the enormous collection of rare books he'd been buying with an abandon that amazed the book world. As he accumulated them they were stored in New York until his new library building could be completed. Then he had them shipped in thousands of crates to California. He also brought to California the library staff he had in New York. The eminent bibliophile George Watson Cole was his director of the library. Since San Marino was almost unbuilt upon open land, the staff had to find accommodations either in Pasadena or our town. Dr. Cole found a place close to the Jacksons' home, and he soon hired young Bill to mow his lawn and do other chores. Impressed by the boy's intellect and his interest in books, he gave him a part-time job on Saturdays helping unpack and distribute the quantities of books that had arrived from New York. Bill spent the year after high school graduation working at the Huntington Library. By that time his future was determined. His great interest was to be early English books, which were the cornerstone of the Huntington collection. He went east to Williams College at the suggestion of Dr. Cole who, at that time, considered the rare book collection in the Chapin Library at Williams to be superior to that of either Harvard or Yale. While still an undergraduate, Jackson began work on the catalogue of the Chapin collection. Later he did the three-volume Pforzheimer catalogue, which is said "provides a standard against which all other bibliographical catalogues must be measured." He became the distinguished director of the Houghton Library at Harvard and at the time of his death in 1964, at age fifty-nine, was America's foremost bibliographer of early English books.

Bill was smart enough to skip a couple of classes and so was in high school when we organized our most ambitious project as kids. As eighth graders we were probably getting "too big for our britches," as our teacher often reminded us. We were restless, ready to enter high school, and impatient. I started circulating surreptitiously around the class a newsletter that I laboriously typed on my brother's machine at home. It was so well received that Pat Kelley and I conceived the idea of a regular weekly printed publication. We arranged with Walter Abbott, a generous local printer, to give us credit. We sold a few ads to local merchants and as joint editors published the first issue of the *Marengo Literary Leader* on October 7, 1919. Pat and I wrote most of the copy for the first issue. He did a story called "The Half Moon," and I wrote a wordy column called "V. Percival Buantee's Corner." On Friday afternoon we set up a stand in the back of the schoolyard and, having warned everyone to bring a nickel to school that day, had a very successful enterprise going. It was too successful. After several weeks, during which we had expanded from two to four pages, the school principal, Mrs. Ballard, decided to appropriate our private enterprise and make it a school project. It was a bitter financial

9

Marengo Literary Leader

MARENGO LITERARY LEADER

Volume I | South Pasadena, Calif., December 2, 1919 | Number 7

THE PURPLE DRAGON

By Lawrence Powell

Taun Fung was pleased. The steamer, Princess Siang Nu, arrievd in the evening with a secret shipment of opium for him. It was to be thrown overboard in cans. Sing Lee, his assistant, would gather them up and bring them to wharf 10. There another servant would receive them and then into his (Taun's) hands they would come. Opium smuggling was quite profitable for Taun Fung. There were ready customers for the dope, all over Frisco Chinatown.

On board the Princess Siang Nu deep silence prevailed. Several shadowy forms stole cautiously across the deck. They lowered what seemed to be large cans of tea over the side into the waiting hands of Sing Lee. Back they stole to their staterooms while the steamer plowed on its way. Sing Lee piloted his little launch into the harbor to wharf 10. Two autos were waiting there to convey the opium to Taun Fung.

As the autos started away a man sprang from behind a pile of boxes and leaped on the back of one of the autos. Through the mystic streets of Chinatown they wound. As they neared the den of Taun's, the man jumped off and scurried for shelter. He ran along in the deep shadows of the buildings until he came to an alley. Down this he went for a short distance then he dodged into a door and climbed two flights of rickety stairs. On the second floor he mounted a small ladder and found himself on the roof. He then walked two hundred paces south and eight paces east. According to calculations he was directly over Taun Fung's secret dope storeroom. Prying a board loose he heard the sound of voices confering in Chinese. He could see into the room through a small crack in the plaster. He recognized Taun's voice but the rest were strange.

The room was fixed in a most resplendent Oriental style. The odor of incense was heavy in the air. Many Chinamen lay in the bunks with pipes of poppy weed in their mouths. Great piles of opium cans were piled in all corners. Suddenly the lid to one of the cans wos raised, out jumped a man with a brace of automatics levelled at the startled Chinamen. At the same instant the man on the roof dropped through the plaster with another pair of revolvers in his hands.

"Who are you that intrudes in the domain of the Purple Dragon?" said Taun Fung.

"The Messrs. Patrick and Donald Sherman of the Department of Justice," they both replied in one breath.

"Who betrayed us"?, shrieked Tuan.

"You betrayed your self," replied Don Sherman.

"I will feed your corpses to the buzzards," howled Tuan Fung.

"Wait until you get them Chinkie old sport," laughed Pat.

Suddenly Don and Pat felt themselves siezed from behind.

"Bind them," commanded Tuan.

"What shall we do with these American swine?" asked Sing Lee.

"They shall be sacrificed to the purple drogon", replied Tuan Fung.

The room was sud[...] to darkness. Pat an[...] bonds cut and they [...] They ran down a lon[...] at intervals by sma[...] walls. The lights bli[...] and Don and Pat [...] falling downward. [...] small boat with [...] flashed before their[...] lapsed into unconsci[...]

When they came[...] themselves in stater[...] room was dimly ligh[...] ple light. The wal[...] oned with large drag[...] dragon was belching[...] sweet smelling incen[...] rocking the boat sli[...]

opened and in came Tuan Fung, Sing Lee and several other Chinamen. Taun Fung did not notice that Don and Pat were awake. "The Sacred and Mystic Order of the Purple Dragon has met here to decide the fate of our two enemies." He brought forth a small bottle of gluey liquid, which was opium. He then inserted a needle into the bottle and brought forth a small drop on the end. Then holding it over a candle flame he turned it slowly around until it was very hot. He brought out his dope pipe and put the opium in the bowl. The other's pipes were soon ready. Then lying down on a couch he said, "Ho!, Nang Yang bring forth the ancient volume of laws of the Sacred and Mystic Order of the Purple Dragon." When this was done he said, "Our most noble ancestors here say that all persons caught having evil intents toward the Purple Dragon in their heathenish brow shall depart by the Purple Scimter. Nang Yang go fetch the Scimter." Nang Yang returned bearing a magnificent scimter with an ivory handle inlaid with emeralds, saphires, diamonds, rubies and many other precious gems. The blade was a wonderful purple steel crescent. "Sing Lee, bring out the prisoners," Tuan ordered. Don and Pat were brought out of the bunks and were placed on a table. "I shall confer the honor upon myself to dispatch them."

Ritchie and Powell in high school

blow, but we did have the satisfaction of being the founders and also of having launched several literary careers, including that of Lawrence Clark Powell, who wrote a couple of short stories that we published.

Larry Powell has been my closest friend during a long lifetime for both of us. Our friendship began early when we started school together back in 1911. As we walked to school a half-mile through the orange groves, our paths would merge before we arrived at the schoolyard. He was assured, aggressive and at that time seemed huge to me. I have since often referred to him as "the tyrant of the teeter-totter," which he was. Our friendship was more completely cemented during our senior year in high school, when I lived with the Powells and became almost another son to the gracious Gertrude Powell. My father had moved the family again, back to Los Angeles. I was loath to transfer to a new school at that time since I was so involved in school activities in South Pasadena. I was on the football team, captain of the track team, president of the Boys' League, and editor of both the school paper and the annual. It was a stimulating year for me. I had been somewhat of a drudge, and Larry enlivened my social life while I influenced his study habits so that for the first time he made the honor roll.

After graduating, Larry and I, along with many other good friends, enrolled in nearby Occidental College in the fall of 1924.

Upon entering college, the question of a choice of a lifetime profession emerges. Unlike Bill Jackson, whose choice had already been made, Larry's and mine were in limbo. At that time we hardly anticipated the direction the years would ultimately take us. I had arbitrarily decided upon law; Larry, who was adept at both the piano and saxophone and had been playing in several dance orchestras, was considering music. We were certainly unaware at the time how much influence a couple of professors in the English department would have on our eventual choice. It was not through suggestion nor recommendation, but they made literature so interesting and exciting, we became devoted to books.

Benjamin Franklin Stelter was the erudite, fatherly head of the English department at Occidental. In our freshman year he took an interest in Larry and me that continued beyond college days. We could always count on him for sage advice. I well remember his college classes. He sat like Socrates, prodding our minds with questions to which we were expected to respond with creative answers. When we questioned him, it was always a revelation of accumulated knowledge from the great intellects. In answering a question in class he would dig into that remarkable memory of his and reply in the words of Plato, Aquinas, Rabelais, or Shakespeare, whichever might be the most appropriate. As a result, he was never self-creative. His thoughts were others' thoughts. This helped make him a great and stimulating teacher but never a creator. His only literary achievement was a concordance of the works of Robert Browning.

Carlyle Ferren MacIntyre was of a different mold, both as a man and as a teacher. The two had been fraternity brothers at the University of Southern California, where, as a faculty member, Stelter had nurtured the promising talent of the younger MacIntyre. And it was Stelter who had been instrumental in bringing MacIntyre to Occidental College after he had received his doctor's degree from the University of Marburg in Germany. Mac was a Villonesque character, a rascal in many ways, a seducer of coeds, a lyric poet, and a writer of risque limericks, one of which has become a classic:

> While Titian was mixing rose-madder,
> His model posed nude on the ladder.
> Her position to Titian,
> Suggested coition,
> So he climbed up the ladder and had'er.

But, with all of his eccentricities, MacIntyre was a great teacher — exciting, stimulating, tough. He was tall, gaunt, and rakish. He raced around town in a sporty Pierce Arrow roadster, always wearing a black fedora askew on his head and carrying a riding whip, which he claimed was "to ward off mad dogs and irate husbands." He'd occasionally pick up some of us after class and drive us into town in his roadster to attack the bookstores along Sixth Street. Having studied in Europe, he'd learned the art of barter. He would sweep through a bookstore gathering an armful of books, offer a ridiculous sum for the whole stack, and then haggle with the proprietor until he'd get them for a bargain price. And in passing he'd pick up a special book, thrusting it into one of our hands while saying, "You buy that and read it."

He lived in a rugged stone house that he'd built himself with boulders gathered from a nearby wash. His nearest neighbor up the slope was an Italian vineyardist, who managed in those Prohibition days to crush a few of his grapes into a hearty red wine, much of which managed to flow down the hill into MacIntyre's snug retreat. Larry and I, along with a number of coeds, were the only ones privileged to share this bonanza. He didn't completely trust the girls, and lest they might find their way to his tryst uninvited, he always blindfolded them when he drove them from the campus to his hideaway. He wrote lovely lyrics for these girls, almost as a diary, which he kept in several large scrapbooks. He'd occasionally let me look through them, which was a very special privilege. He excited my imagination and urged me to write, especially poems. It is a wonder that he was allowed to stay at this small Christian college for four years. Finally President Bird was able to trade him to UCLA for a professor in disfavor there. They, in turn, shipped him to the Berkeley campus after several years. He was always a great favorite with the students but managed to antagonize members of the faculty.

I jumped around a bit during my college years. After my freshman year at Occidental, I spent four quarters at Stanford, continuing to find stimulating teachers. Edith Mirrielees and Margery Bailey were the two I encountered there. I also spent one quarter at the University of the South before returning to Occidental for my senior year. At Sewanee, as the University of the South was known, I was fortunate in finding two fine teachers in William Knickerbocker and Eugene Kayden.

14

While I was still planning to pursue law as a career, my interests were drifting toward books. After my freshman year at Occidental I worked a few months at Vroman's bookstore in Pasadena before going to Stanford and there began collecting first editions and press books. In 1928 the Huntington Library opened its doors to the public for the first time. It was an important event locally, and I was among those attending the first day. I was particularly attracted to some bindings on display by the Englishman Thomas James Cobden-Sanderson. The concept of binding intrigued me — working with books and being creative at the same time. Not too long after this a classmate, Josephine Hodges, aware of my interest in fine books, arranged for me to visit Alice Millard, a friend of her family and a dealer in rare books.

I had never seen the Millard house, though I had lived in the area most of my life. It was well-hidden in a short, circular lane off Prospect Boulevard. It was awesome to first come upon this spectacular, secluded structure, nestled like an Aztec temple among trees, astride a small gully. I was ushered into this picturesque house to meet Alice Millard, who was to become a dear friend and patron.

When younger, as Alice Parsons, she had studied art in Europe, where she had become interested in the Pre-Raphaelites. On returning home to Chicago she visited McClurg's bookstore, looking for books on the artists. The rare book room in McClurg's, dubbed "The Saints and Sinners Corner" by Eugene Field, was presided over by the stately bookman George Madison Millard. On

Alice Parsons' first visit there, Millard sold her a first edition copy of J. W. Mackail's *The Life of William Morris.* Many visits followed and a romance developed between this twenty-eight-year-old girl and the fifty-five-year-old bookman. They were married in 1901 and were soon looking for a home. A young architect, Frank Lloyd Wright, built "a modern wooden dwelling" for them in nearby Highland Park, where they lived until Millard's retirement in 1914. They traveled extensively in European countries, foraging for books. Alice's knowledge of rare books grew as she became acquainted with many of the most important rare book dealers and collectors in England and on the Continent. Among these were the Cobden-Sandersons and Mae Morris, along with the booksellers Maggs and Quaritch, with whom she had many dealings in later years.

After Millard's retirement they moved west and bought a small bungalow on Huntington Drive in South Pasadena, only a few blocks from where I was then living. Being but nine years old at the time, I was hardly a prospective customer and was never aware of them. They continued their book business, with Alice assuming many of the responsibilities as her husband became feebler. After his death she continued, bringing rare books from England to sell to wealthy California collectors such as Templeton Crocker and William Andrews Clark.

When Frank Lloyd Wright was working on the Imperial Hotel, he'd often stop en route to Tokyo to visit with his friend Alice Millard. She had for some time been contemplating building a new home for herself and her books

16

The Millard House by Frank Lloyd Wright

and had bought a nice conventional lot in Pasadena. Having a comfortable nest egg of $10,000, she asked Wright if he could do a house for her for that amount. He agreed he could but didn't approve of the property she had bought. Searching the area, he finally found a vacant piece that appealed to him. It was part of a small ravine that emptied into the nearby Arroyo Seco. The irregularities of the property appealed to him visually, and he designed a unique house straddling the dry ravine. Construction was a nightmare for both Alice Millard and Wright. His supervision from afar, as he spent most of his time in Japan on the Imperial Hotel, led to many costly mistakes and agonizing labor problems. The cost ran far above the original budget and pitted a strong, opinionated woman against a man who wanted to do things only his way.

When it was completed, however, it was an architectural marvel reminiscent of an Aztec temple. Wright remarked as he viewed the finished building, "I would rather have built this little house than St. Peter's in Rome." It was constructed of concrete blocks designed by Wright with a decorative cross motif. For some of the walls the motif was hollowed through, with glass sandwiched in the middle. The effect, when the sun shone through the blocks, was a dazzling light pattern on the opposite wall. When the fluttering leaves of the trees outside deflected the light, the images on the wall danced in delight in continuing playfulness.

To build in a ravine in southern California invites disaster. While it seldom rains, when it does the lowlands are flooded and the hills become rivers of mud. The Millard house built in this natural ravine acted as a dam when the first serious rain came. A lake formed before the water finally burst through the lower level of the house, following its customary course down to the Arroyo. Wright had overlooked including a master bedroom in his plan, an oversight that had also escaped the watchful eye of Mrs. Millard. She was forced to convert the library into a semi-bedroom where she slept on a daybed. The flat roof leaked through the years despite every effort to control it. One feature that often bothered me was a low concrete beam that ran above the entrance to the stairway to the upper levels. It was just high enough not to bother the petite Alice Millard but with anyone taller it required a foresighted duck or else a bruised forehead.

On my first visit I was not only dazzled by the house but by its mistress. Alice Millard was small and delicate, belying her drive and intensity. She weighed only a hundred pounds but it was a hundred pounds of sustained energy. She was born in 1873, so must have been fifty-nine or sixty years old at that time. She was attractive, but in her intensity she would tighten her lips and bare her teeth as she talked, which could make one feel somewhat uncomfortable. She always wore blue to match her piercing eyes and her blue-tinted white hair which was her crowning glory. She was an animated talker, especially when discussing her favorite books, which included the Kelmscott Chaucer and the Doves Press Bible. I had already been fascinated by the Cobden-Sanderson bindings I'd seen at the Huntington Library, and her enthusiasm for his beautiful books added to my interest in printing and binding.

Despite my growing interest in the art of the book, after graduation I enrolled in law school at the University of Southern California. I was not particularly happy. Law did not seem to satisfy the urge for creativity that had been gradually developing during my experiences in college. USC was across town from where I lived, and I'd usually stop on the way home to visit and chat at the various bookstores that lined Sixth Street and Grand Avenue. Dawson's Book Shop was always one of my favorite haunts. Ernest Dawson, whom we called "Father Dawson," imported great quantities of books from England through Marks & Co. of London. He made the merchandise look always new and different by continually shuffling the books around. Since he didn't like having the same books around for too long, he marked the date of purchase and the cost in code on the flyleaf of each book, and if it hadn't sold within a few months, he'd reduce the price in half. If it didn't sell then, he'd reduce it again. His more impecunious customers, such as I was, would play a waiting game to see when a certain book would come within their budget. Often we were disappointed, as some other collector who might not be as patient would purchase the prize before it had been sufficiently reduced. It also involved us in a tantalizing game of hide-and-seek because of Mr. Dawson's penchant for rearranging the books. Here I became acquainted with such noble book collectors and historians as Robert Ernest Cowan, Henry Raup Wagner, J. Gregg Layne, Carl Wheat, W.W. Robinson, Phil Townsend Hanna, Frederick W. Hodge, Paul Jordan-Smith, and Robert Woods.

Around the corner on Sixth Street was a variety of small bookshops intermingled with some shoddy hotels. I'd spend most of my time at Jake Zeitlin's "At the Sign of the Grasshopper," where there were always examples of the newer books from the great private presses such as Nonesuch, Golden Cockerell, Merrymount, Grabhorn, and Nash, and designers like Bruce Rogers and Eric Gill. The shop was also the hangout of many young writers, artists, and printers who were both stimulating and congenial.

I knew very little about the man Thomas James Cobden-Sanderson beyond having seen his beautiful bindings and a few examples of the books from his Doves Press. Thus, when I spotted a copy of his *Journals* marked down from $25.00 to $6.25, the temptation was so great, I bought the two volumes and hurried home to read about the man. That slight happenstance completely altered my life's plans. I read how Cobden-Sanderson had abandoned a successful career in law at the age of forty-five to learn the art of bookbinding. With his example in mind, I decided to give up the study of law and find a way to become a bookbinder or a maker of beautiful books.

Hardly knowing where to begin and looking for guidance, I went to the Huntington Library, where I knew many of the staff. Robert Schad, the curator of rare books, introduced me to a couple of the young pages who distributed the books in the protected stacks. These were Gregg Anderson and Roland Baughman. They had printed several small things under the imprint of the Greybow Press at Gregg's uncle's print shop in nearby

Pasadena. Schad thought they might be able to advise me, and so they did, showing me some of the things they had printed, along with examples of the student work being done at the Laboratory Press at Carnegie Tech under the supervision of Porter Garnett. They suggested I should first try San Francisco, where the finest printing on the West Coast was being done, and that I should seek advice and possibly an apprenticeship there. I was soon off to San Francisco to see the two finest printers there. I first went to the Grabhorn Press and met the brothers Ed and Bob. I got much advice, and friendships that lasted until their deaths many decades later, but not a job. Next I saw John Henry Nash. He was somewhat self-important and pompous, but extremely helpful. In his printer's apron (as he had been setting type), he took time to show me his impressive library of the great printed books of the past centuries. I was greatly impressed, not only with the books he showed me but with the greater-than-lifesize painting of himself that graced one wall. Since I didn't know a pica from a quoin, he suggested that I should return home and get some local instruction before I sought apprenticeship with him. He referred me to his friend Bruce McCallister in Los Angeles. McCallister was a fine commercial printer who was just beginning to get interested in fine printing through his membership in the Zamorano Club. He arranged for me to be admitted into the printing course at Frank Wiggins Trade School, now called Los Angeles Trade Technical School.

I found myself in a class of thirteen and fourteen year olds. They, as well as the teachers, were somewhat puzzled by having an older college graduate wanting to join them in this elementary beginning printing class. But fate was generous. My printing instructor was a tall, soft-spoken man named James Halleck. When he heard my name was Ritchie, he asked if I'd come from Michigan. I hadn't, but my parents had come from a small town there called Imlay City. He smiled in recognition and told me that he had grown up on a farm near that town, and when he first started school there he was a frightened farm boy coming to the "big city," though the town had barely five hundred inhabitants at that time. The town boys began picking on this little farm boy until the town banker's daughter intervened and became his guardian. That little girl was my mother's younger sister, and her good deed a half-century earlier made my own somewhat awkward start in a new school easy and pleasant. Instead of having to take all the elementary instruction of the younger students, I was allowed to create my own curriculum.

I had already been reading everything I could find about fine printing and was not interested in the usual exercises required of beginners. I wanted to print books, and the opportunity came almost before I had learned how to set a line of type. I stopped by Jake Zeitlin's shop to tell him I was now a printer. He was pleased and asked if I'd like to print a poem by Carl Sandburg. I was thrilled at the prospect of starting my printing career with such a noble author and readily agreed. Jake had come to California from Texas only a few years earlier. In Fort Worth, while working on the newspaper, he had interviewed Sandburg and subsequently accompanied Sandburg's readings with his guitar. They had become friends, and Sandburg had written a short introduction for Jake's first book of poems printed by the Grabhorn Press in 1927. When I had come to tell Jake I was now a printer, he had just received a letter from Sandburg, including a recent poem, *Soo Line Sonata.*

It took me at least a week to set the type, as I was just beginning to learn the arrangement of the type case and in which slot each letter was kept. As with most beginners, I wanted to decorate the page, as I assumed that was the function of a designer. I slightly overdid it, and when I took proofs proudly to show Jake, he gently suggested we abandon the project. His decision was partly influenced by a following letter he'd received from Sandburg saying he'd prefer him to do another poem. I was disheartened after all of my work and expectations, but upon reappraising what I had done, I realized how bizarre my over-decoration must have appeared to Jake. I removed the decorations, reproofed the type, and showed it again to Jake. He was generous in his approval this time and suggested that since I'd put in so much work, I should print a few proof copies for him. And thus I began my career as a printer with this now very rare booklet called *Soo Line Sonata.* Many years later, when I was asked to give the Engelhard Lecture at the Library of Congress on "Fine Printing: the Los Angeles Tradition," it was my first visit to Washington since 1946 when I designed some books for the Government Printing Office. Peter van Wingen graciously escorted me on a tour of the Library and showed me displayed my printing of *Soo Line Sonata.*

24

SOO LINE SONATA

By
CARL SANDBURG

1929

SOO LINE SONATA

HE WAS SAYING,
I have looked deep
into the cisterns of the stars
and they were all strugglers.

He was saying, I hunted alligators
on the moon once
and I saw they had excellent teeth
even as the camels there
were humped with excellent humps,
single humps, double humps.

He said he saw a milk white horse
on the peak of an iceberg
looking for a place to slide down
to the pearl purple sea foam.

After and before

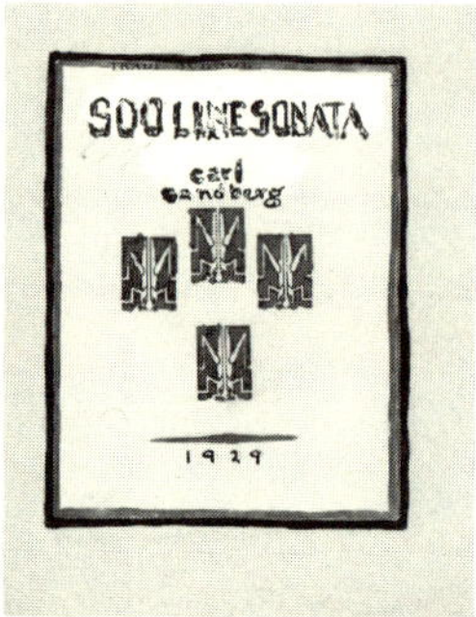

25

He was saying,
I have looked deep
into the cisterns of the stars
and they were all strugglers.

He was saying, I hunted alligators
on the moon once
and I saw they had excellent teeth
even as the camels there
were humped with excellent humps,
single humps, double humps.

He said he saw a milk white horse
on the peak of an iceberg
looking for a place to slide down
to the pearl purple sea foam.

He said, I've been lonely,
but never so lonely
one wind wouldn't take me home
to the four winds.

I have seen broken hearts
in little piles like egg shells,
dogs with their heads
laid on crossed fore-paws,
women asking the simple question,
"Who's going to close these dying eyes?"

Yes, since you ask me,
I've been lonely,
but never so lonely
one wind wouldn't take me home
to the four winds.

Yet never once
have I known time to fail me,
time with its dark doorways,
its monotonous mumble,
its steady tattoo of dust to dust,
ashes to ashes.

Open those doors — shut those doors,
slip those bracelets on,
take 'em off —
I want to stay here,
I want to go home —
What do you care? What do I care?

I have nailed up a slogan:
Blessed are they
who expect nothing,
for they shall not be disappointed.

I shall be less faithless, I guess,
to a line of vanished verse:
Be not rash with thy mouth.

Yes, he was saying,
I shall take Ezekiel for a partner,
use the proud assumption
of the Apostle Paul,
borrow the manner of the men
who go down to the sea in ships,
correct the wanderers
who do not know that love is costly,
please myself with the laughter
of Babylonian babblers at their best,

and join the tears of her
whose pitcher was broken
at the well.

I shall sit weeping with the fellow
who told his boon companions,
"Get me drunk
and have some fun with me"
and the Iowa farmer
who said of a horse,
"She was stylish
but she couldn't stand grief."

And when they ask me,
Whose fool are you? I shall answer,
your fool, my fool, Gawd's fool —
and if they go on asking,
What would you do
rather than anything else?
I shall offer the comeback,
Go up in a chariot of fire like Elijah.

Yes, he was saying,
I shall slide down
into pearl purple sea foam
into cisterns of stars.
I shall be lonely,
though never so lonely
one wind won't take me home
to the four winds.

So he was saying.

I'm not aware if Jake ever showed or gave a copy to Sandburg, or if it was ever again published in any of Sandburg's books of poetry. However, Jake had mentioned to Sandburg that he'd like to print something of his and not long after this Jake gave me another Sandburg poem to print. It was called *M'Liss and Louie.* I printed a hundred copies of this, which Jake sold for the sum of one dollar each. These were harbingers of Jake's venture into publishing and mine into printing.

At that time I was also attending a Saturday class in contemporary poetry being given by Hildegarde Flanner at the Pasadena Playhouse. Since I was now a printer, I decided I'd print some keepsakes for those in the class, giving me an opportunity to practice my new art as well as pleasing my classmates. I wrote to several poets whose verse I admired and received permission from most of them to print one or two of their poems. While waiting to hear from them, I practiced on some of my own, which I had written in college. I hid my identity under a variety of pseudonyms. The first of these was *The Slough of Despond* by James Beattie Pitwood, followed by *Dream in a Garden* by Betsy Ann Bristol. Fortunately, I began getting replies from some of the real poets I had written and could offer some better material. These booklets I proudly presented included *Women* by Louise Bogan, *Interrogate the Stones* by Archibald MacLeish, *Valley Quail* by Hildegarde Flanner, and *Midsummer* by Léonie Adams. By June of 1929 my semester at Frank Wiggins was complete and Hildegarde Flanner's classes were over. I still had a couple of sonnets by Robinson Jeffers to print with nowhere to do them.

While in college I had come to know the amiable and eccentric printer Clyde Browne, who handled most of the college printing. His "Abbey of San Encino" was a few miles from the college in an area he called "Garvanza Olde Towne." He had laboriously built his "abbey" over many years, hauling boulders from the nearby Arroyo Seco. A picturesque building backed against a hillside, it included a chapel, a dungeon, living quarters around a cloister, and a printing shop with a large stained glass window representation of a monk watching an Indian operating an old wooden printing press. Up the hill, in back of the printing wing, Browne had built a series of small studios, which he rented to artists and writers. I had taken a job at Vroman's bookstore but wanted to continue printing. Powell and I had rented one of these studios, and I arranged with Clyde Browne to use his presses on Sundays for the sum of one dollar. I had bought a couple of cases of used type from Frank Wiggins. I moved these into the studio and again began printing — setting type on evenings and printing on Sundays.

My first project was the printing of the two sonnets by Robinson Jeffers, which I had planned for Hildegarde Flanner's class. It was somewhat more ambitious than the previous booklets inasmuch as, with Clyde Browne's help, I hard-bound it. Proudly, when it was finished, I sent copies to Jeffers and distributed a number of others. Few people seem to bother to read such small gifts. Powell was the exception in this case and gleefully pointed out several misspelled words that no one else appeared to have noticed. It was a good lesson in bibliophilism for me. I immediately corrected and reprinted the small book. The corrected edition now can be bought for a reasonable sum, but the faulted first printing sells for many thousands of dollars.

The year I worked at Vroman's was a busy one. As I try to remember those days, I wonder where I found the time to print. The bookstore hours were from seven-thirty in the morning until seven-thirty at night, and on Saturday until ten o'clock. But I somehow managed to spend enough time at the studio and printing shop to produce quite a number of items. Several I did for Robert Cowan, the erudite and genial librarian of the William Andrews Clark Memorial Library. He was "Sir Robert" to his intimate friends, "Mr. Cowan" to others. He was of the old school, fastidious with an extraordinary memory. His speech was precise and his knowledge of books prodigious. I printed several things for him, including the most ambitious project I'd thus far attempted, a booklet called *The Booklover's Litany,* which he presented to his fellow members of the Zamorano Club. I also continued my "errored" way and misspelled the club's name in the colophon. To this day, no one seems to have noticed it. The ever-observant Powell was not a member of the club at that time and, not being a recipient of a copy, was not able to further malign my spelling lapses.

I enjoyed doing one little booklet of a poem that he had unearthed and suggested to me. It was called *Beer* and had been written in the last century by an obscure English poet named George Arnold. I love the lilt of its opening stanza:

Here,
    With my beer
I sit,
    While golden moments flit:
    Alas!
    They pass
Unheeded by
And, as they fly,
I,
Being dry,
    Sit, idly sipping here
My beer.

I also printed a book of a couple dozen poems by my former professor, Carlyle MacIntyre. He had allowed me to skim through the scrapbooks of his accumulated poems and choose a selection. My choice may not have included many of his best lyrics, but I was fortunately able to preserve a few of the hundreds of his early lyrics. A few years later his home and all its contents were lost in the Montrose flood of 1933. He never regained the youthful exuberance of those early lyrics, engaging most of his later years in translations of the works of French, German, and Spanish poets.

While working at Vroman's, I spent most of my meager salary on books, especially those dealing with printing and printers. Among them was a copy of volume four of *The Fleuron,* a magnificent English publication on printing. There was an article on "The Future of Printing," in which I read that the books of the Frenchman François-

32

The Schmied Atelier

Louis Schmied "may be considered as the most exact anticipations up till now of the book of the future." There were several illustrations in the article of intriguing pages from his books. Shortly thereafter I saw a copy of his *Le Cantique des Cantiques,* which Mrs. Millard had recently purchased in Paris. It was the most exciting book I had ever seen, and I decided I must study with this man. When my mother and aunt planned a trip to Europe early in 1930, I tagged along and thus was able to get to Paris.

With unwarranted confidence and youthful brashness I managed to find the *atelier* of François-Louis Schmied on the obscure Rue Hallé near the Port d'Orléans and bravely knocked at the door. My deliberate college French seemed to be incomprehensible to the young fellow who answered my knock, but he did gather that I wished to see M. Schmied and finally led me up a flight of stairs to a reception room. I nervously sat upon the comfortable lounge while looking around the room. On one wall there was a case of books that I presumed Schmied had designed and printed. There were also numerous sleek modern pieces of sculpture that I learned had been done by his friend Gustave Miklos. In time a tall, athletic man emerged through a side door and began speaking to me in rapid French. I couldn't understand him and replied in the slow and deliberate French I had been taught. Evidently he couldn't understand me either, as he quickly turned about and exited by the door he had only just entered. I sat wondering what was going on, whether I had been summarily dismissed, or whether he was getting someone who could decipher my gibberish. I sat and sat and sat, all the time getting more apprehensive. It is very possible that this initial lack of communication was the key to my eventual career. If he had known at the moment of my intentions, he probably would have dismissed me abruptly. But I had aroused his curiosity and he returned eventually with the son of his friend Jean Dunand, a young man who spoke tolerable English. He immediately asked me what I wanted of M. Schmied. I answered that I had come all the way from California to work for him. There was a short period of apprehensive silence on M. Schmied's part and then the question to me, "Where is California?" I answered that it is where Hollywood is. That must have been the magic word, for in those days Hollywood was magic. After some discussion, Schmied shrugged his shoulders and left. Young Bernard Dunand turned to me saying, "If you have come all the way from Hollywood just to work for M. Schmied, he feels he can hardly send you back. Start to work on Monday."

Before I left for France, Hildegarde Flanner had suggested that I look up her sister Janet, who was writing the biweekly "Paris Letter" for the *New Yorker* under the name Gênet. She lived in a small Left Bank hotel on Rue Bonaparte. It was named the Hotel St. Germain des Près after the historic church nearby on the corner of the Boulevard St. Germain. It was hardly a luxury hostelry, but the rates were reasonable, about thirty dollars a month. I moved into the garret room, five floors up the steep, winding stairs. It had recently been occupied by Henry Miller, who had deserted New York for a fling at Paris. He wrote a friend: *I love it here, I want to stay forever . . . I will write here, I will live quietly and quite alone. And each day I will see a little more of Paris, study it, learn it as I would a book. It is worth the effort. The streets sing, the stones talk. The houses drip history, glory, romance.* He couldn't long afford the luxury of this wonderful cubicle and I happily inherited it. I could sit looking over the rooftops at the lofty steeples of the church while watching the activity on the street. On the floor below me lived Janet Flanner with her friend Solita Solano. They lived in comparative luxury, having converted a couple of rooms into comfortable quarters, even installing a private toilet and bath. The rest of us had to share a stand-up toilet, one for each floor, with a skimpy supply of torn newspaper pages hung on a spike to be used as TP. There was only one bath, except for Janet's, for the whole hotel. Fortunately, it was handily located on my floor. It was seldom used except by a few Americans. It was necessary to arrange with the *garçon* to haul up a few pails of warm water for which he charged a franc. He was a friendly lad and seemed perplexed at my frequent baths, explaining that most of his friends bathed only once a year and then would take a girl along.

Janet Flanner took a sisterly interest in me. She kept me supplied with books to read and would occasionally take me out to eat at a restaurant somewhat more expensive than the ones where I usually ate. On one occasion she allowed me to write part of her "Paris Letter" about an exhibit of Schmied's books at the Galeries Georges Petit.

Life for me in Paris in those days was an enriching experience. Powell had spent a few weeks with me before going on to Dijon and the university there. Together we explored most of the well-known café haunts. Les Deux Magots, Le Flore, and Brasserie Lipp were just around the corner from my hotel. Most of the tourist action in those days was on the Boulevard Montparnasse, where the Dome, Rotunde, Select, and other cafés were bustling most of the night. We investigated most of them, often in the company of a couple of California girls we had met. After Powell's departure I settled down somewhat. Schmied had taken me under his wing, and I lunched each day with him and his son and daughter in the studio. Often at night I'd buy myself a long spike of tasty, fresh French bread and a snip of butter, which they'd shave with a thread of wire from a beehive-looking mound, and together with a bottle of wine and some runny *Port du Salut* cheese, I'd have a gourmet feast in my room. Other times I'd grab a meal at Figons, a small restaurant on the Boulevard St. Germain where for six francs one could get a small steak of questionable pedigree, *pinards,* and a carafe of red wine, also of questionable pedigree. On more affluent occasions there might be a *choucroute garni* dinner with a mug of good Alsatian beer at Lipps. Lipps was also my favorite haunt for later in the evening when most of the Americans working on the *Herald-Tribune* would gather there for companionship. Paris cafés were poor men's clubs — a place to gather for an evening with one's friends. I'd often see James Joyce at one of the back tables of the Deux Magots or Picasso at La Flore.

After my return home I wrote nostalgically to Powell, still in Dijon: *I should like to haunt the bookstalls again, or experi-*

37

Les Deux Magots in 1930

Ritchie Groenewegen Powell Paris Summer 1930

ence those warm summer nights when I'd walk or sit on the river's edge with Marge, that beautiful American girl, with the lights of Printemps making gay colors on the water and the prow of the Isle dark and menacing with its hangover of history. Great God, how I loved the tall tower on the church of St. Germain des Près. Even the subway there was nice on cold winter days. Each morning I'd ride it to Schmied's. In the winter months there were roasted chestnuts at the corner down the street a block or so, and I'd pluck a bottle of wine to drink with them from the little grocer across the street on Bonaparte. The Tuileries in Spring, with the tulips. And at night with its pink lights. It is a little section of the world that I once knew well and I know I'll never recapture it as I knew it. How true this observation was to be. When I returned thirty years later I found my once quiet neighborhood teeming with gawking tourists congregating at the Deux Magots and Flore. Lines of people were in front of the Brasserie Lipp waiting to be seated. My little hotel had had a face lift and had been modernized. Montparnasse was no longer the reigning, bustling center of activity. The era of the twenties and thirties was history.

The mid-twenties were the glamour days of Hemingway, Fitzgerald, and the score of wild American expatriates, but there was enough glamour left in 1930 to excite a youth. Elliot Paul, Robert McAlmon, Gertrude Stein, Nancy Cunard, and Sylvia Beach were still there, prolonging the fading glory of the twenties. I was happy to be a witness and a small part of this twilight era, but I was actually more interested in the vital printing being done at the time. Numerous Americans were involved, including McAlmon, who published Hemingway's first book as well as Gertrude Stein's massive *Making of an American.* Sylvia Beach published Joyce's *Ulysses,* using the French printer Maurice Darantière, as had McAlmon. Nancy Cunard, Bill Bird, Edward Titus, Harry and Caresse Crosby all were publishing contemporary literature. I paid passing attention to their activities because I was primarily interested in the great French printers of that time. The artist-printer I worked for was the most important, as time has proved, but I visited with others such as Louis Jou and Maurice Darantière and watched them at work. Schmied also encouraged younger artists to illustrate books, and he set up an auxiliary plant across the street from his *atelier* called Les Éditions du Bélier, which published books illustrated by young artists, including his son Théo.

Dijon was also a magnet for much of my attention. Powell was there studying. He had been lured there in part by the Fishers, who had preceded him and were the only other Americans then attending the university. Mary Frances Kennedy and her sister Anne had been students at Occidental College. M.F., as we called her, left school to marry Alfred Young Fisher, a brilliant and handsome young scholar who was bound for France to pursue studies for a doctor's degree in literature. Powell was seeking the same.

Having these friends in Dijon, I would make frequent weekend trips there to enjoy their company, along with the appetizing Burgundian fare. The Fishers had arrived there a year earlier and had found a comfortable *pension* in the home of the Rigoulots at 14, Rue du Petit-Potet. Mme. Rigoulot's father, whom we called Papa Zie, was

40

Fisher and Powell at the Foire Gastronomique Dijon 1931

Alfred Young Fisher, M.F.K. Fisher, Lawrence Clark Powell on vacation at Cros de Cagnes in April 1932.

reputed to have been the finest confectioner in France, and the tasty creations he often served us on Sundays readily confirmed that judgment. Upon his daughter's marriage to Rigoulot, he had settled upon the couple a large dowry that Rigoulot had managed to squander. This explained why they were reduced to taking in boarders, but they still continued to wine and dine with the lavishness of their better days. M.F.K. Fisher learned much about gastronomy and wines while living there. We experienced the *Foire Gastronomique* with M. Rigoulot when he was selecting a year's supply of those great Burgundian wines for his cellar. While he sipped, judged, and spat, we gulped — and suffered the next day, to M. Rigoulot's amusement.

We were all young and full of the confidence of youth. Larry was writing the "Great American Novel," which died somewhere as his career took a different turn. I was writing artificial poems while Fisher, the most mature of us, was completing an epic poem of enormous proportions. M.F., who subsequently became the most successful writer of us all, usually sat silently by as we would gather to read to one another our latest effusions. She kept a diary, and much of what she wrote about food is quite autobiographical, stemming from these early days.

Each day, when school and the library were closed for the lunch break, Al Fisher would retreat to the Café de Paris to watch the parade of trollops and to write. On sunny days, which seldom came to Dijon, he'd sit on the terrace. On other days he'd find a table inside, where he'd cover pages of his notebook with the poetic words that flowed endlessly from his mind. In time he covered more than a thousand pages in his miniscule hand. It could have continued for many thousand pages more had not M.F. decided to return home to visit her parents. With her departure Fisher's muse retired, and he never wrote another word of his epic. Powell later wrote of that summer how he and Fisher were "together day and night, walking, talking, studying, drinking, smoking, a relationship like the one Ritchie and I knew the year he lived in our home." He went on to say: *Fisher and I drank our noon* apéritifs *at the café on the Place du Theatre, along the route taken by the prostitutes on their weekly visit to the medical inspector, and we would banter with the girls as they passed. "When are you coming to see us?" they asked. One night we did, making the rounds of several whorehouses, buying a few drinks at each one but no more. At the last stop I was high. Sitting down at the piano in the parlor, I began to play and never stopped until daylight showed at the window. The girls were enchanted. Fisher danced with each in turn, and they ended up by giving us breakfast. The Madame offered me a job, in return for free meals. I declined.*

I happened to be visiting in Dijon on Armistice Day in 1930, which we celebrated by going for a drink at the Café de Paris. We had long noticed a sign over the bar offering "Cocktail Americaine." None of us had ever tried it. In honor of the day, and as we were the only Americans in Dijon, the proprietor generously offered us a free cocktail, which was an unusual occurrence. As he joined us at the table, I noticed he had only a glass of wine for himself. We toasted with the "Cocktail Americaine." It was an experience not to be repeated. After expressing our hypocritical appreciation, we asked where he'd learned the secret of mixing such a drink. He explained that during World War I a group of American soldiers had invaded his bar asking for a cocktail. Not knowing how to mix one, he'd allowed them to do it themselves while he watched. They had put practically everything they could find into a shaker and shaken it well. He had remembered how they had done it, but this was the first time since then that he had an opportunity to mix one. He was very pleased we liked it. Some thirty years later when I revisited the Café, the sign had disappeared. I didn't have the temerity to ask for another "Cocktail Americaine."

On each of my visits to Dijon, the four of us would splurge for one triumphal dinner at the finest restaurant in the town, Aux Trois Faisans. It was almost a ritual dinner for us, hardly ever varying. It was a bit of nostalgia for home. We'd start with a salad and a carafe of the house Chablis, which came chilled and was always good. Then we'd have a blood-rare *chateaubriand* with *haricots verts* and a bottle of beautiful ruby-red Chambertin which we could hardly afford, but it was an occasion. Many years later M.F. returned, hoping to recapture the excitement of dining as we'd known it in those youthful years of ours. She wrote in one of her books of once again climbing the stairs into the restaurant, looking for the many familiar features. She asked to be seated where she could be served by the waiter who had been so indulgent and helpful when we were such innocent gourmets. He was older now and somewhat inept, spilling a bit of wine and shuffling hesitantly, but with a remembering smile.

My life in Paris was equally absorbing. I did many odd jobs around the *atelier,* including setting type for trial pages, running errands, helping with the bottling of Schmied's private wine, and teaching English to his daughter Madeleine, or Maddy as we called her. I was quite attracted to her, and Schmied seemed to approve of my interest, inasmuch as he had brought her into Paris from their country home to have me teach her English and look after her in general. When I wrote her some love poems, however, she admonished me, saying it was improper for one to write poems to a girl who was engaged. Meanwhile, Schmied would tell me of his displeasure with her choice, saying he was a businessman and he would much prefer to have her marry an artist. It wasn't until after I had left Paris that I learned Maddy had given some consideration to me. My brother had saved a clipping from the *Los Angeles Herald* headlined "Mystery Death of Beauty Probed." It told of the mysterious death of a young French dancer who had been appearing in Hollywood and San Francisco. Among her possessions were found several letters, including one from Madeleine from Paris expressing hope that she was finding happiness in America and inquiring for information concerning a young man known as Ward Ritchie. Madeleine said she was interested in Ritchie and wished to learn more of his "social grade." By the time I was aware of this interest, I was far away and she had married her businessman.

My arrangement with Schmied was fairly flexible, and I would decamp from time to time to explore and to see other printers. I visited the Plantin Museum in Antwerp and the Klingspor Foundry in Germany. I had earlier met young Karl Hermann Klingspor during his wandering days when he spent a few months at the Grabhorn Press in San Francisco. Fritz Kreidel was there, but I was disappointed on finding Rudolf Koch away on vacation, since I had so admired his work. Another time I took a more prolonged vacation. The climate in Paris had been miserable that winter. It was clammy, cold, and there was never even a glimpse of blue sky for months on end. For a Californian that could be miserable, and I yearned to see the sun again. I heard there was an island paradise in the Mediterranean as warm and beautiful as California, and I managed to get to Mallorca and finally warm myself. It was as nice as I had anticipated and luxuriously inexpensive. Alfonso XIII had recently been deposed, and the peseta was so cheap that rooms in the best hotels were not more than two dollars a day. I hadn't been in Palma long before I ran into John Cage in a bookstore. He had been a roommate of Gregg Anderson's at Pomona College and had earlier contacted me in Paris. He was young then, perhaps eighteen or nineteen, and had come abroad to study music or architecture, he wasn't sure which. After a few months of puttering at each, he decided that he'd prefer to explore the islands of the Mediterranean and had finally arrived in Mallorca after having visited Capri and other islands. He had picked up with a young Harvard student and they had found a large house in nearby Pollensa, which they had rented for about twenty-five dollars a month, including a servant woman. He suggested that I might join them if I'd pay a third of the rent.

John had heard there was a small English press across the island at Deyá. Our cook's brother had a car, and John and I arranged for him to drive us to Deyá one warm sunny day. It was a lovely little town high above the blue sea. We lunched on the porch of a quaint café with a bottle of the worst wine I've ever tasted and began inquiring about the whereabouts of the elusive English press. At each house where we inquired, we added a few more curious followers to our retinue. A German I questioned gave me the first clue, answering my query with, "You mean Graves' Press?" Then I realized it must be the Seizin Press of Robert Graves and Laura Riding about which I had read.

With the retinue of a few dozen villagers, we finally found the house. It was perched high above the road on a bank. I climbed the stairs and pounded on the door with no answer. Descending, I was almost ready to give up the search when there appeared on the bank above us a woman in full rage. She stood glowering down at our group of stragglers, her sturdy legs spread wide in anger, her hands on her hips, her hair wild in the wind. She shouted down at us, "What do you want?" Quite cowed by her appearance, I meekly looked up at her and replied, "We are looking for the English press that we understand is here in Deyá." She calmed down, dismissed our followers, and led us to her study in the back of the house. Seated finally at her desk, her first question was, "Do you know who I am?" Fortunately, I had read Robert Graves' autobiography, *Goodbye to All That,* and realized that this must be Laura Riding — and so I responded.

47
Laura and Robert

Laura

Robert

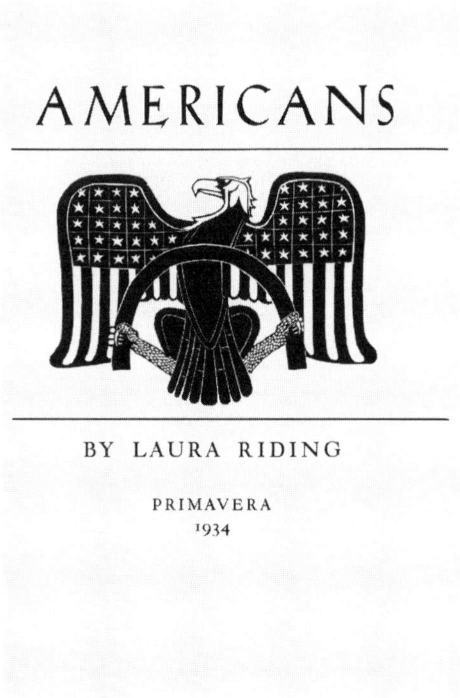

AMERICANS

BY LAURA RIDING

PRIMAVERA
1934

She immediately relaxed and we talked about the press, the bits they were printing, where they got their paper, and the printer in Palma who set most of their type. She told us that so many came only to seek out Graves that she had become annoyed. Only the previous week she had thrown out Liam O'Flaherty, but my inquiry had been so frank and specific — about the press — that she had allowed us a visit. Besides, I had known who she was and had not asked to see "the great man." After a short time we left her study and wandered into the house to see their handpress and some of their books. I asked her if she enjoyed printing. She answered, "For me it's not fun. It is just plain work."

Robert Graves eventually came up from the beach where he had been sunning. We had a few glasses of sherry (as I recall, Laura Riding drank only tea). Though I never saw either of them again, we corresponded for many years. As many of us were, Laura was saddened by the death of Hart Crane. When she mentioned she was planning to write an elegy for him, I told her I'd like to print it. It was never written, however, and instead she sent me the manuscript of a poem called *Americans.* I printed it, and sometime later she sent me a copy of *Focus,* an informal little magazine she and Robert occasionally produced for their friends.

In it Laura Riding had written: *The only thing that has happened from America has been the publication of my ill-tempered deliberately shabby little poem,* Americans, *which I wrote a year ago for a rather nice young man called Ward Ritchie, who has a press in California. He printed it beautifully, with forthright red adorn-*

ments, and now I am feeling somewhat shamefaced — on his account, not the poem's — because, without any suggestion of a whine, he has written to say how much he liked the Leaves, *which I sent him, and how much he would have enjoyed being their printer. It was through Ward Ritchie, by the way, that the whole German situation arose here. He came to the village about four years ago, looking for the Seizin Press, mentioning my name and Robert's, and met the German called Herpes, who said, "Oh, you mean Graves' Press." When he found us he was surprised to see me, having got the impression that I was no longer concerned. Which, of course, started a thing with Herpes, who said, "If a woman expects personal recognition, she shouldn't live with a man in the same house." This went on to other subsequent bitter remarks, until one angry night Robert strode up to the village and into the café and slapped old Herpes' face. Whereupon all the Germans in Deyá were infected with a strong injured German–colony disease which has passed on from one season's German colony to the next in true post-war spirit.* Little had I realized the consequence of my short visit to this charming seaport village.

Before I returned home I visited England for a couple of months. When I arrived there the spring flowers were beginning to bloom and I was able to enjoy the resurgence of life in the gardens and vast parks of London. But I was more interested in books than flowers, and after finding fairly inexpensive lodgings at the Penn Club in Tavistock Square, I began making the rounds of the bookstores. One of the first I visited was Putnams. I was waited on by a very pleasant girl who had been somewhat disfigured by an accident. I asked to see what Nonesuch Press books they might have. It brought a smile to the girl's face as I told her of my admiration of Nonesuch books and Francis Meynell, the man who was responsible for them. I added that I hoped I could meet him while I was in England. She said, "If you would really like to meet him, please come to dinner tomorrow night. He is my uncle, and he and his wife, Vera, will be with us. My name is Sylvia Lucas."

I realized that this was the girl D.H. Lawrence had written about in *England My England.* The Lawrences had stayed for some six months in one of the cottages in the Meynell family compound at Greatham in Sussex. The author had based his story on the Meynell and Lucas families, telling of how little Sylvia at the age of six had fallen on a sickle, which had disfigured her for life. I enjoyed her warmth and intelligence. In addition to working at Putnams, she helped edit some of the Nonesuch books.

Needless to say, I accepted her invitation and arrived a bit early the next night at the Lucas home at 47 Palace Court. I was entertained by Sylvia's pretty sister and her cousin Wilfred before Sylvia appeared. We were soon joined by Wilfred's younger brother and sister. They had lived in Santa Barbara in California for several years, where their father, Everard, had gone to recuperate from tuberculosis. Assuming that all Californians played tennis, they hoped that I would join them during the summer at their place in Sussex, where they needed a fourth for their game. I was scheduled to leave for home on the fifteenth of June and had to decline.

Francis and Vera Meynell arrived. He was tall and lean, a trifle bald, with a good-natured wit. She was lovely and quite silent. After dinner we climbed the stairs to the table tennis room and we all batted the little celluloid ball around. I thought I was fairly good, but Meynell clobbered me. I later learned that he was quite an athlete, not only a skilled table tennis champion but also captain of a cricket team, and he had been an outstanding soccer player in school.

He invited me to visit him at the Nonesuch Press, which I did a few days later. I took with me some sample pages from books designed and printed by Schmied. I also presented him with a copy of Robinson Jeffers' *Apology for Bad Dreams* that I had printed on Schmied's Stanhope handpress. Meynell called in E. McKnight Kauffer to see these things. Kauffer had his studio in the same complex and had illustrated several books for Meynell. He told me that he had been born in Montana and had worked for four years with bookseller Paul Elder in San Francisco. "San Francisco," he said, "perched above its beautiful bay, is the one place in the world for which I feel nostalgia." He was intrigued by the concept, the variety, and the technique of Schmied's work. Meynell, on the other hand, considered them as artworks and outside the province of the typographical book. He mentioned that he was interested in the text of books and wished to publish books that, because of their length, frightened the average publisher. This he eventually did with his compact editions of Donne, Blake, and William Morris. He considered his problem as one of condensation, while Schmied's concept called for expansion of the text to accommodate his decorations and illustrations.

After we had chatted awhile, we descended to the

ground floor to see the composing room. Meynell designed all of the Nonesuch books but had them printed at various printing plants. His equipment consisted of a small proof press and, as he phrased it, "a thimbleful of a great variety of type faces." There was a single compositor working there who set trial pages following Meynell's instructions. His method of creating a page intrigued me. It would be set in type, made up, and a proof pulled. This would be pinned up on a blank wall. Meynell would study it there, suggest a change to the compositor, and a new proof would be pulled to pin up beside the first one. This would continue with many variations in type, color, and concept. I was amazed to see almost forty variations of a title page crowded onto the wall. To me, at that time, any one of them would have been acceptable, even great, but Meynell seemed determined to explore every alternative before making a final decision. When I returned to California and set up my own shop, I reserved a blank whitewashed wall to use in the same way. During the early years when I was printing, I must have walked a hundred miles between the Washington handpress on which I proofed and that wall.

Americans too often think of the English as especially aloof and unfriendly. My experience has been the opposite, that they are extremely friendly and helpful. I had little more to offer when I visited there than my enthusiasm for books and printing, yet dozens of people went out of their way to be helpful. A bookseller by the name of Evans, whose shop was across from the British Museum, closed his place one day to take me to the Curwen Press and introduce me to Oliver Simon. This great printer was generous in giving me examples of his work, while discouraging me about my future plans. He told me that the "private press" as such was dead and that the United States was overrun with such presses. He advised that the best path to follow was the one Daniel Berkeley Updike had taken — the production of fine commercial work.

Cobden-Sanderson

At dinner, a few days later, with the Richard Cobden-Sandersons at their home on the bank of the Thames in Hammersmith, in the building where his father's Doves Bindery had been located, Cobden-Sanderson concurred that the day of the "private press" had passed. He doubted if his father could have succeeded with his Doves Press in this year of 1931. The house, however, at that time was a treasure trove of Doves' memorabilia, much of which has now come to California through the efforts of Alice Millard. Many of T.J. Cobden-Sanderson's personal bindings and the great pattern books of all his bindings are at the Huntington Library or in the private library of Norman Strouse, given to the University of California. Later, when I was at Cambridge and Sir Sidney Cockerell was showing me around the Fitzwilliam Museum, I spoke enthusiastically about the Cobden-Sanderson things I had seen. He brusquely said, "He threw Emery Walker's type into the Thames." That was that. At the time I was unaware of the acrimony that had resulted from the event.

A most pleasant visit was to Shelley House, where I had been invited for tea by C.H. St. John Hornby and his wife. I arrived at their home on Chelsea Embankment and was waiting at the gate when I was joined by another young fellow who had also been invited for tea. He introduced himself as Philip Hofer, then of the New York Public Library. We were led upstairs to the Hornbys' library, where after a few minutes of conversation we were served tea by Mrs. Hornby. During the tea I had an opportunity to look over the lovely library. There were bookcases all around the room. Later we were shown some of the treasures they contained, mostly early illuminated manuscripts in immaculate condition. Mr. Hornby told us he would never consider buying a book not in fine condition. Many manuscripts were also displayed on several tables, together with a large collection of inlaid and decorated boxes. Hung high on one wall over the bookcase was the oar he had used when his crew had won the Oxford-Cambridge race.

After tea we walked out through the garden to a small building that housed the Ashendene Press. It was a tidy little printing shop we saw there. They were printing the *Thucydides* at the time on an Albion handpress. I asked Hornby if he were still helping with the printing. He replied that he only read proofs and checked the sheets nowadays, but for many years he had set the type. In order to keep his hands respectable for his visits to his office at W.H. Smith & Son, he had covered them with a preparation called Peldow, which he could wash off after printing, leaving his hands white and clean. He mentioned that, in 1894, when he had been working in the printing shop of the Smith firm, of which he was now president, a friend took him to visit the Kelmscott Press. They were printing the Chaucer. He had a friendly talk with William Morris

55

Ashendene Press

From C. H. StJ. Hornby, Shelley House,
Chelsea Embankment, London, S.W.3.

19 May 1931

Dear Mr Ritchie

If you could come here to tea at about 5.15 on Wednesday, May 27 I should be pleased to see you and show you my Press. I am afraid that that is the only time I have available during the next 10 days.

Yours truly

C H St J Hornby.

and was so impressed with the operation that he decided to have a press of his own. He bought an Albion handpress similar to the ones he had seen at Kelmscott and moved it into a shed in his father's place in Hertfordshire. Ashendene, the name of the estate, was also the name he selected for his press. He acknowledged that his first productions were rather wretched, but he was then quite proud of them. He sent a copy of one to William Morris, who thanked him graciously and said it was nicely done. Hornby remembered this encouragement with gratitude.

This may account for his graciousness to me when I later sent samples of my printing. When he wrote me that he was closing the press because his pressman was getting too old to continue and he doubted if he could train a new one, I remembered his saying he might sometime have one of his presses available if I should want it. I wrote to ask if I could now buy one, but he replied he still couldn't bear to part with any of his Ashendene Press equipment. He died in 1946 at the age of seventy-eight. It is ironic that not long after his death, Dawson's Book Shop in Los Angeles received a large crate from Marks & Co. of London. As was usual, Mark Cohen would ship off to Dawson's from time to time his overstock and materials that he thought Dawson might be interested in selling. This crate of odd materials had been picked up at a country auction and, probably without even looking at it, Cohen had shipped it off to Los Angeles. When the box was opened at Dawson's, it was found to be filled with wood engravings and loose printed sheets. An astute clerk recognized them as being from the Ashendene Press. They were quickly sold at modest prices and are now scattered.

John Johnson of the Oxford University Press was one of the most interesting men I was to meet. My recollection of him at the time was of his being reasonably young, bulging a bit about the middle, with chestnut hair as fine as silk, which hung over his forehead in the manner of Aubrey Beardsley. He had been collecting ephemeral printing for years — bits and pieces that seldom survive but can show much about the habits and manners of the decades in which they were printed. While his enormous collection is now in the Bodleian Library, it was then kept at the press, where he could watch, enjoy, and organize it. After I returned to the United States, I sent him many small items and corresponded with him regularly until his death. Generously he once wrote: *We are a rather stern factory of nearly 900 souls, and to very few is given the privilege of wandering in and out of their own free will. But Bruce Rogers has that privilege. Come to see the rather stern factory, and become the second person who has the privilege of coming in and out at his own pleasure.*

The war years were particularly trying ones for Johnson. In one letter he wrote: *Already some 250 of our younger men have gone off to the wars, and daily and insensibly we pass through transition from manpower to womanpower, under the not ungracious tutelage of the trade unions. The trade unions are facing their responsibilities splendidly. And all the time even the nature of our printing changes, not I hope the quality of it.* Later, he wrote that he never left the premises of the press. *I still sleep on my little camp bed, surrounded by a historic library, in the same room in which I work.* It wasn't until some thirty years later, when I visited the John Johnson Collection in the Bodleian, that I learned that during the daytime of those war years the University Press operated as usual, but at night Johnson and a dozen so-called "fire wardens" printed secret war material. I then understood why he would never leave the premises.

Francis Meynell had invited me to the thirtieth meeting of the Double Crown Club, which had been founded a few years earlier by a group of bookmen and printers. St. John Hornby was the speaker that evening, and almost every English printer of importance was there. I was introduced to Bruce Rogers, a visiting American printer, by Francis Meynell, who had evidently told him of my apprenticeship with Schmied. Rogers said he was acquainted with Schmied's work, having seen the exhibition of his books at the Seligman Gallery in New York in 1927. While he wasn't impressed by Schmied's typography, he was impressed by his color work. He mentioned that to him the most interesting printer of all time was Erhard Ratdolt, and if he could have afforded it he would have collected all of the books Ratdolt had printed in the fifteenth century. He invited me to visit him in the office Emery Walker had put at his disposal in Clifford's Inn.

Our meeting there was a "show and tell" affair. First I showed him some Schmied pages. He thought the work was too perfect. In fact, in his opinion, the French hadn't done any really good books since the eighteenth century. He then began to show me some of his things, and we looked and talked at length about books and printing. The first thing he showed me were pages from T.E. Lawrence's translation of the *Odyssey,* on which he was working. He said that he enjoyed his association with Lawrence and was saving the letters he had received from him with the idea of eventually printing them. We also looked at many of the proof pages of the lectern Bible then being printed at the Oxford University Press. The pages were plain and forthright as the book was designed primarily for being used on the lecterns of the churches in England. Oxford's supply of their previous printing was running short, and a new printing was required. The Press wrote Rogers for permission to use the Centaur type that he had designed. While considering this request, he decided he'd like to revise some of the letters. During the discussion he was asked if he would be willing to design the new Bible for them. He agreed, asking only as a fee to have fifty copies for himself to be printed on handmade paper in a larger format with more ample margins. The regular church edition, while printed from the same type, was planned to be on machine-made paper in a smaller format. The Oxford Press was delighted with this arrangement. As the work progressed, Oxford asked Rogers if he would mind if they printed an additional fifty copies of the limited edition for themselves. He agreed to this and to their later request to increase this number. Eventually a couple of hundred were printed. This book, one of the outstanding printed books of the twentieth century, was completed during the years of the Great Depression, and sales were disappointing. Years later I had a letter from Rogers telling me that he still had a few of his fifty copies left, hoping that I could induce the Huntington Library to buy a copy from him. He mentioned that the Oxford

University Press also had copies available. He showed me the designs he had made for the *Book of Common Prayer* in the competition with Daniel Berkeley Updike and the Oxford and Cambridge University presses. Updike's was chosen, a more sedate and less decorative concept than Rogers'.

Rogers went on to tell me that one of the best books printed in the nineteenth century was the Pickering edition of Euclid. It was also a book that he would like to do. In general, he didn't care much for illustrated books, though he did like Gordon Craig's illustrations for the Cranach Press *Hamlet.* He didn't care too much for Rudolph Koch nor Eric Gill as illustrators. He liked Gill as a friend and as a sculptor, but he thought his illustrations too monotonous. He showed me several layouts of books he would like to do but probably never would. One was the Euclid and another was *Aesop's Fables* with woodcut reproductions from a fifteenth-century edition. This latter book, however, he thought he might do for the Limited Editions Club. He had tried to steer clear of the Club, but George Macy had been so very considerate recently and had sent announcements of the *Odyssey* to the Club's membership, so he felt obligated to do the book. Subsequently he did several, including the many-volumed Shakespeare. I understand that he never liked the illustrations done for each play by different renowned artists and had removed the illustrations from his own set, leaving only his version of the text. Macy allowed him to design the two volumes of poems as he wished, without illustrations and with simple ornamentation of printer's flowers.

61

Apology for Bad Dreams

The time finally arrived for me to return home. I had seen a number of the great printers of Europe and had acquired experience and numerous books in the process. My days in the *atelier* of Schmied had been exciting. I had done almost everything there, from learning a bit of wood engraving to bottling wine. I had even printed a book. After I had printed Robinson Jeffers' *Stars* earlier in California, I had asked him if I could have some more of his poems to print. He suggested several that had appeared in *A Miscellany of American Poetry 1927.* I brought them with me to Paris and in leisure time printed them on Schmied's Stanhope handpress. I designed them with some suggestions from Schmied, and printed an edition of thirty copies of *Apology for Bad Dreams.* I proudly gave copies to Schmied, to Théo, and to several of Théo's friends who had included me in their activities. Among these was Jacques Chenais, a dapper young fellow of whom I had become quite fond. He, along with Théo and Claude Laurens, the son of the sculptor Henri, and I, with a few others, would gather together on many merry occasions planning to revolutionize the world of art. With the passing years, Théo continued printing in the tradition of his father but without quite the same genius. Claude was a successful sculptor but never as famous as his father. Jacques, on the other hand, became world-famous as the premier puppeteer of our time. He carved magnificent puppets, wrote plays for them, and managed their shows throughout Europe and even for a command performance at the Smithsonian Institution in Washington.

On my way home to California, I stopped in New

York to seek out the printers there whose work I had admired. Among these was Elmer Adler of the Pynson Printers. Schmied had given me a copy of his *Le Livre de la Vérité de Parole* when I left, together with a bound set of various leaves from other of his books. These, together with my *Apology for Bad Dreams,* I showed to Adler in the library of his immaculate printing shop located on one of the upper floors of the *Times* building. In addition to many fine books, he edited and published *The Colophon,* a hardbound quarterly in which each article was designed and printed by a different printer. He selected these from among the most renowned in Europe and America. The magazine was, thus, valuable both bibliographically and typographically.

My Jeffers book immediately interested Adler. He had been unsuccessful in getting Jeffers to write an article for him on the publication of his first book, *Flagons and Apples,* as part of a series he was running on first books of well-known authors. He asked me if I might have some influence. While I had printed some Jeffers poems and corresponded with him, I had never met him, but I told Adler I'd try to help.

When I returned to California, the first thing I did was to buy a Washington handpress, for which I paid fifty dollars. I had previously purchased a couple of cases of Goudy's Garamont type from Frank Wiggins Trade School. That was the extent of my printing equipment when I drove north to see Jeffers and my friends Gloria and Gordon Newell, who were also living in Carmel. Gordon and I had been fraternity brothers, along with

Larry Powell, at Occidental College. He was now an embryonic sculptor and married to the young actress Gloria Stuart. They knew the Jefferses, and the three of us drove down to the Jeffers home, Tor House, on Carmel Point to see Robin and Una. They had been happy with the copies of *Apology for Bad Dreams* that I had sent them, and we talked somewhat about Powell, who was writing his thesis at Dijon on Jeffers' poetry. When I brought up the matter of his writing the article on his first book for *The Colophon,* he told me that as he recalled there was nothing of interest to write about. When I returned home to southern California, I began snooping around to see if I could find anything of interest about *Flagons and Apples.* It had been printed by the Grafton Publishing Company at Jeffers' expense. Five hundred copies had been printed, with many love poems to a married woman, Una Kuster. His effort to sell copies met with litttle success, but it hardly mattered at the time, because his Una was divorcing Kuster and she and Robin had fled together to Seattle to be married. The printer was left with almost all of the five hundred copies and wanted to dispose of them. He wrote Jeffers, and Jeffers told him to junk them. Instead, he sold them to the Holmes Book Company, where the books sat around for several years. Finally, when the store was being moved from Spring Street to a new location on Sixth Street, Holmes, in an attempt to dispose of as much stock as possible before the move, had an enormous sale of books. He set up a stand in front of his store from which he'd shout inducements to passersby to come into the store and march out with great bargains. From time to time, to attract further attention, he'd throw out a handful of books into the street for onlookers to scramble for. In this way he got rid of most of his stock of *Flagons and Apples.*

After the publication of Jeffers' *Tamar* in 1925 and, subsequently, *The Roan Stallion, Tamar and Other Poems,* Jeffers became a favorite among collectors of first editions. One day as Holmes was poking around in the basement of his store, he came upon a stack of the long-forgotten *Flagons and Apples.* He took a few copies and placed them on one of the tables that lined the front of his shop, with a simple sign that priced the copies at $2.50. Shortly afterward he noticed that they were gone. He put out another stack priced at $5.00. They too quickly sold. This continued, with the price spiraling up to $25.00 before his supply was exhausted. He then rued his earlier excesses with copies. This story, which I got from Mr. Holmes, I passed on to Jeffers with some other material I had unearthed. It must have prompted him, because not long after I heard from Adler that Jeffers had sent him his story of his first book, of which he was sending me a copy.

While I had been up north seeing Jeffers, I had also visited San Francisco to see the Grabhorn Brothers. Ed sold me some battered Eve type designed by Rudolf Koch, and from Hazel Dreis, who was then binding the Grabhorn *Leaves of Grass,* I bought a ream of Arnold unbleached paper to add to my meagre printing equipment. With what type and equipment I had, I sat up most of the night setting Jeffers' story. The next morning I proofed it on the handpress using the paper I had from Hazel Dreis. I

FIRST BOOK : ROBINSON JEFFERS

65

Jeffers' Hawk Tower by Landacre

I AM WILLING to tell the history of my first book, though it is not clear why it should interest anyone; certainly it does not interest me. In 1912 I came into possession of a little money, a little more than was immediately required, a novel ex-

sketched in a semblance of Jeffers' Hawk Tower, where he did his writing, and sent the proofs off to Adler. He must have been taken by surprise, but evidently he liked what I had done, because he wrote to ask if I had the equipment to print an edition of 3,000 copies for him. I could hardly tell him that I had only a handpress, so I deviated slightly from the truth by writing him that I had a Colt's Armory Press, such as I'd seen the Grabhorns use in printing their books. When Adler wrote to say the job was mine, it became necessary to quickly find a press. Fortunately I was able to find a similar type press that was being discarded by the local print shop that years earlier had printed the *Marengo Literary Leader* for me. I got some nice paper from Blake, Moffit & Towne, which they were closing out. Paul Landacre cut a fine wood engraving of Hawk Tower for me, and the first production of the Ward Ritchie Press appeared as the opening article in Part Ten of *The Colophon,* along with such internationally known printers as The Oxford University Press of England, R. & R. Clark of Scotland, Klingspor of Germany, and Updike's Merrymount Press of the United States. One could hardly wish to start in better company.

Gradually I was able to add more equipment and fix up the old place in back of the family home. I built a fireplace in one corner with a gleaming copper hood, bricked in a patio garden, and even had curtains on the windows, which one of my female friends contributed. I was doing books, catalogues, and letterheads, mostly for the area bookstores — Dawson's, Zeitlin's, and Alice Millard. Alice was a tough and exacting patron, but I loved doing her announcements and catalogues of the occasional exhibits she put on in the Little Museum of La Miniatura, which she had built on the lower level of her property. The most ambitious of her catalogues was one of an exhibition of Doves Press books and Cobden-Sanderson bindings. I cut on wood a long-stemmed letter "P" to open the text, reminiscent of the spectacular "I" with which Cobden-Sanderson had commenced the Doves Bible. The great calligrapher Edward Johnson, when he saw my "P," graciously criticized it as having the weight of the strokes misplaced. It was a problem getting Mrs. Millard to release the proofs for final printing, inasmuch as she was always waiting for additional books to include. Finally, a day or so before the exhibition was to open, we had to start printing, regardless. My pressman and I were working almost continuously, still hoping somehow to get copies ready for the opening, when on the evening preceding the reception, as we were weary-eyed printing sheets, the whole building seemed ready to turn over. It was the severe earthquake that practically destroyed the nearby city of Long Beach. I didn't dare stop and incur the great wrath of Mrs. Millard, so all through the night we printed, accompanied by earthquake aftershocks. I hadn't slept for thirty-six hours, but managed to finish the printing and the binding of a hundred copies, which I delivered to Mrs. Millard just as the first guests arrived. She hardly recognized me with my red eyes and a couple of days of beard, unwashed and in dirty printer's clothes. But she was relieved and happy to see the catalogue. She later revised it and I printed another edition on Dard Hunter paper.

I worked in a strange way. While there was a phone in the house, I didn't have an extension in my little shop in the rear. After my mother had died early in 1933, it made it somewhat difficult, but instead of getting another phone, I made almost daily calls on my customers, mainly the bookstores huddled around Sixth and Grand in Los Angeles. It also gave me the opportunity to pick up books, which were dirt cheap in those days. This took care of most of my days, and I'd print at night if I were lucky enough to get any jobs on my wanderings. Occasionally I'd wander as far as Hollywood to visit the Satyr bookshop or Stanley Rose's little shop, a haunt of many of the movie writers. On more fashionable Wilshire Boulevard was the Penguin Bookshop, run by Louis Samuels. It was away from the hub of the bookstores, and there wasn't much drop-in trade. When customers did happen in, Louis would hover over them and deluge them with books to peruse. He lived in a secluded glen in Eagle Rock with his artist wife, Grace Marion Brown. It was a suburb close to South Pasadena where I lived, and I'd often drop in for a visit.

On one occasion he had a very merry party. A wary customer had wandered into his shop and, as usual, Louis gathered a batch of books for him to look over. Among the books was a copy of Melville's *Moby Dick,* illustrated by Rockwell Kent. As Louis effusively praised the illustrations, the gentleman modestly admitted that he was the artist. Samuels was so pleased, he immediately asked Mr. Kent if he might be free for the evening to join with several friends at his home for cocktails. When Kent accepted, Louis got on the phone and invited some cronies and customers to join him later at his home to meet Rockwell Kent. He closed shop, visited his favorite bootlegger to get a supply of gin and whiskey, and drove Kent to the hideaway in Eagle Rock to await an influx of revelers. It was a convivial evening, with everyone enjoying the party, including Kent. For some reason one of the guests, Jake Zeitlin, became suspicious. Though he had never met Kent, he'd had an exhibit of Kent's prints in the small gallery in his bookshop. He sneaked into another room and called a girl who had been intimate with Kent and asked her for a description. When she described him as not too tall but balding, Zeitlin was certain that this short, hirsute guest must be an imposter. Jake kept mum, not wishing to spoil the party. Later Louis asked Jake if he'd mind driving the guest back to Hollywood where he was staying. On the way there Jake accused him of being a fake. The only reply from the man was, "What impertinence." They drove the rest of the way to Hollywood without another word.

After this impersonation was disclosed, it was found that the false Rockwell Kent had once been a groom on the estate of the real Rockwell Kent at Ausable Forks in New York. His name was Michael Gergeson. Before working briefly for Kent, he had been a hanger-on and notorious party-crasher in Greenwich Village, where he had assumed the name of Mike Romanoff and was tolerated as an amusing eccentric and dubbed "The Prince." He later reassumed his "Prince Romanoff" guise and soon became a Hollywood celebrity. Sometime later a

70

Rockwell Kent

Mike Romanoff

handful of Hollywood actors and writers financed a restaurant for him in Beverly Hills, which became famous as Mike Romanoff's.

I had been an admirer of the work of Rockwell Kent even before I started to print. The first press name I had used was The Flame Press, inspired by a Kent engraving I had seen in Zeitlin's shop. It was of a reclining man with an arm upraised against a background of a huge flame. I had also been impressed when I saw the two-volume edition of *The Canterbury Tales,* translated by William van Wyck, with his illustrations. Soon after I had arrived in Paris, Van Wyck invited me to his sumptuous apartment on Boulevard Montparnass and showed me the book, which had just been published. While I regetted not meeting the real Rockwell Kent at Louis Samuels' party, soon afterward I read of a planned lecture tour that would bring Kent to California. He had been a member of my college fraternity, and I wrote him asking if he'd be willing to give the annual Phi Gamma Delta lecture at Occidental College. He agreed, and we filled the college auditorium with an enthusiastic audience of students. He was a spirited speaker, telling wondrous stories of his adventures in remote, mostly frigid areas of the world — Alaska, Newfoundland, Greenland, and Patagonia.

After the lecture a few of us, including Kent, Jake Zeitlin, Larry Powell, and Gordon Newell, gathered at my studio for a bit of food and several gallons of red wine. Kent kept us entertained with stories of his many escapades while in college and after. I still remember one rather amusing tale he told. During World War I, before the United States was involved, he was living and painting in a remote village in Newfoundland. He enjoyed provoking the suspicious villagers, who were curious about his roaming the countryside painting. He added to their suspicions by painting a sign above his cabin door reading "spy outpost." At one time he decided to have a tennis court and leveled off some of the land, only to have his neighbor plow it under. He was furious and threatened to kill him. The neighbor dragged him into court, and the trial was a major event in the village. Nearly everyone attended, and there was little sympathy for Kent. The possibility of acquittal seemed hopeless when he took the stand in his own defense. When asked if he had threatened to kill his neighbor, he answered "Yes." He then went on to tell the jury that after the neighbor had plowed up his tennis court, he was so angry that he had called him every unsavory descriptive invective he could think of. And he continued with a flow of discourteous invectives until the courtroom, including the judge and jury, was convulsed with laughter. He concluded by saying, "I indeed said I'd kill him, cook him, and even eat him if he weren't such a stinking neighbor." The case was laughed out of court. When the United States entered the war, he pulled up stakes and returned home. The uninhibited and articulate Kent was almost a clone of Carlyle MacIntyre.

As the evening wore along and the red wine filled cup after cup, everyone was in a mellow and fraternal mood. Larry Powell had recently returned from France with the thesis he had written on the poetry of Robinson Jeffers. Jake was a bookseller and publisher, I was a printer, and

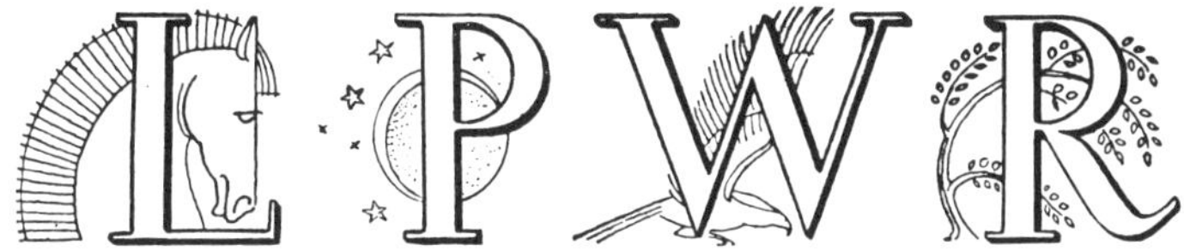

73

Kent Initials

Kent was an illustrator. It was decided we'd do a book together using Powell's text on Robinson Jeffers with Kent illustrations. There was no time lost before I had made a tentative layout, including a hawk to decorate the title page, a large opening initial, and a decorative initial for each of the chapter openings. When Kent asked me what those chapter letters might be, I had no idea, but not wishing to stifle our immediate enthusiasm for the project, I wrote down ten or so letters, including an "L" and a "P" for Powell and a "W" and an "R" for me. Kent stuffed my notes in his pocket and the party continued. The next day I was doubtful if anything would result from our drunken plan, but Kent kept his promise and a few weeks later his drawings arrived. Powell reluctantly rewrote the opening paragraphs of each of his chapters to accommodate my choice of initials. The book was printed as a trade book for the Primavera Press by a commercial printer for economic reasons. It was to my design, but I regretted not being allowed to print it myself.

Later on in the evening Kent mentioned that he had done a series of drawings for Goethe's *Faust,* but the publisher had reneged because of the Depression. He said he would like the book to be published sometime with a suitable English translation. I was interested and suggested to Carlyle MacIntyre that it might be something he'd like to do. He had taken his doctor's degree at the University of Marburg in Germany and was fairly fluent in that language. MacIntyre wavered for a bit, trying out his skill on a few pages. He finally agreed and was given a Guggenheim fellowship to help further his work. When he sent the final manuscript to me to design the book, he included a note, "Well, your little dream of some five or six years ago when you suggested Kent's idea to me (and I God-damned both of you for the hard work) is materializing." Primavera Press had folded by this time and the book was published by New Directions.

I subsequently printed a few small things of Kent's. And in 1969, when he was eighty-seven years old and I was almost ready to retire, I became involved in his final book. It was a reprint of his previously published *Wilderness,* but with additions. When he and his young son had spent a winter on isolated Fox Island in Alaska, he had kept an illustrated diary. When the account was originally published, considerable material was left out. His wish was to have an edition with the complete text. There were complications. Kent, who had Communist sympathies, had given the original manuscript to the Soviet Union, where it was considered a national treasure. It was necessary to have it returned to the Soviet Embassy in Washington, where we had it microfilmed in order to get the complete text as he had originally written it. We printed fifteen hundred copies with all of the deleted material and numerous additional drawings. He had been extremely gratified, when the book was originally published, that an English critic had written that it was "the most remarkable book to have come out of America since Walt Whitman's *Leaves of Grass.*" He was also pleased with our collaboration, writing, "I am delighted that at last a book of mine is to be produced by that most distinguished printer — yourself, hoping that you will share my own enthusiasm for the whole project." We sent the pages of the colophon to Ausable Forks for him to sign. His signature was still firm at his age, but it exhausted him. We were able to get a few copies of the bound book for him to see just before he died on March 13, 1971.

When I started the Ward Ritchie Press in 1932, I was completely naive about business practices. In estimating a job I'd include the cost of materials, underestimate by far my own time, and ignore such things as overhead and profit. When I finished and delivered *The Colophon* job, I was unaware that I was supposed to send along an invoice. I agonized for a couple of months when I wasn't paid, until they wrote and asked for a bill. I was not an aggressive salesman, but my friends at the bookstores sent me both their business and an occasional customer. Jake Zeitlin was a particularly good friend. He was only a few years older than I but quite experienced in the ways of the world. He was a catalyst and drew to his shop the younger writers, artists, and printers of the area. He'd come from Texas a few years before, stone-broke, with only his guitar and a bagful of books. He had hitchhiked, leaving his wife and baby daughter behind until he could find employment to bring them on. He was penniless and almost starving when he arrived in Los Angeles. He washed dishes in a local cafeteria just to get something to eat. He next found a job shoveling manure in the Doheny gardens before finding more congenial work in a series of local bookstores. Finally he felt confident enough to light out on his own. With a satchelful of books, most of which he had on consignment from Dawson's Book Shop, he'd peddle to reference libraries in the movie studios and to the few collectors he'd come to know.

I was still in college when I first came to know him. He had converted what had once been a back entrance to a building on Hope Street into a shop with an appropriate address for his aspirations. It was only a narrow hall, and probably the smallest and most cramped bookshop in existence. He prospered there and was soon able to move around the corner onto "booksellers row," along with a dozen or so other bookshops that then lined a few blocks of Sixth Street. His new shop was still small but attractively designed by Lloyd Wright, the son of Frank Lloyd Wright. It included a small gallery of hardly more than a bare wall, where he had exhibits of the work of Edward Weston, Paul Landacre, Rockwell Kent, and Kathe Kollwitz, among many others.

Jake had an abiding interest in publishing books as well as selling them. Following his experience with the two small Sandburg items I had printed for him, he published a controversial talk that Merle Armitage had given to the California Art Club. It was entitled *The Aristocracy of Art.* Armitage was forthright in his opinions, saying, "Arriving in California from New York, I was appalled by the anemic, colorless art being produced in the then remote and provincial Los Angeles area. Moral, social, and sentimental values were utterly confused with aesthetic values." This concept appealed to the group of young iconoclasts who gathered at Jake's shop, and they joined in publishing a periodical, *Opinion.* Jake was the editor, and the whole group offered their opinions for him to include. It lasted for only a year when either the members began to disagree or else they ran out of opinions.

Also, Bruce McCallister, the premier commercial printer in Los Angeles at that time, was getting the urge to publish books. As a commission he had printed *The History of Warner's Ranch.* Young Grant Dahlstrom, recently employed by him, had helped design the book. It was the first book from Los Angeles to be selected as one of the Fifty Books of the Year by the American Institute of Graphic Arts. This was an added stimulus to the growing local interest in producing books. A wealthy book enthusiast, Marguerite Eyer Wilbur, had translated from the German an early book about Los Angeles called *Eine Blume aus dem Goldenen Lande oder Los Angeles,* published originally in Prague in 1878. McCallister and Jake published it as a joint venture.

While occasional books had been printed in Los Angeles previously, there had never been what could be called a true publisher in the area. At that time, in 1930, there were eager authors around, many of whom were willing to pay to see their writing in print. To remedy the void, Jake formed the Primavera Press, named after the Spanish word for spring. Lacking capital, Jake started his new venture pretty much as a vanity press, doing subsidized books by local authors. McCallister was enlisted to print most of the early books published by the press. He also published under his own imprint the handsome *California Hills,* a book of Paul Landacre's wood engravings. McCallister's firm became a victim of the Depression, and Jake turned to me to print the next Primavera books.

Shortly before I left England in 1931 to return home, I received a cablegram from Jake offering me a job in his bookshop at $100 a month. While I was still intent upon being a printer, I needed time to get started, and this job would give me a bit of interim income. I spent my first fifty dollars buying my Washington handpress. A couple of my college classmates, Edward Hackett and Paul Newell, had started a small printing shop and asked me to join them as a silent partner while I continued to work with Jake. Jake was agreeable for the time being, but after a couple of months he concluded that I might become a better printer than a bookseller. He sent me packing with commissions to print a catalogue and a book for him. The book was *Libros Californianos or Five Feet of California Books,* compiled by Phil Townsend Hanna, the scholarly editor of *Westways.* It was the first nonsubsidized book the Primavera Press published. Soon after the book's publication, the firm of Hackett, Newell and Ritchie succumbed. Fortunately for me, the commission from *The Colophon* arrived soon after, and the Ward Ritchie Press was foaled.

It was during my brief employment with Jake that the Rounce & Coffin Club was founded. On one November day, near closing time at the shop, a couple of young printers, Gregg Anderson and Grant Dahlstrom, were still hanging around looking at press books, but not buying. Jake suggested that he'd like to close up shop, but if interested the four of us could go out and have a bite to eat together. The Zamorano Club had been formed three years earlier by some of the more affluent book collectors, and we felt left out and decided that evening to form our own club. It was a completely zany organization during its early years, with no purpose other than conviviality. We would take turns having the monthly meetings at one another's homes, with the poor wives expected to feed us. Powell catches the flavor of the group in *Ten Years (Almost) of Rounce & Coffin-ism,* an early short history he wrote. *The origins of the Rounce & Coffin Club are already shrouded in mystery. Only one fact is irrefutable: it was founded by an itinerant book-peddler and a band of starveling printers — Jake Zeitlin, abetted by Gregg Anderson, Grant Dahlstrom, and Ward Ritchie — back in the bitter autumn of 1931. Thus it would seem to have been the product of social unrest and unemployment. Or, in other words, the materialized figment of a group of disordered young brains, all of them hungry for business. For Satan finds some mischief. This little band, augmented very early by Saul Marks and Paul Landacre, met at first — disguised as idealists on idealism bent — in Zeitlin's den, across the street from the Bible Institute. Those of you who recall the gatherings in that evil sink (Zeitlin's shop, that is), with its sordid trappings of cash-drawer, wrapping-paper, twine, and even a few grimy books, will be amazed that the present dignified organization could have metamorphosed from such an unpromising sabbat. And a glance at the early minutes of the club (that's all you'll be permitted, anyway, because fear of blackmail has caused them to be buried deep in the archives) will reveal the depths of bibulous obfuscation and versification the orgiasts could and did descend.*

The club finally found a purpose and settled down into a sober and respectable organization. For some time it had been felt that the American Institute of Graphic Arts, in its yearly selection of the Fifty Books of the Year, was too East Coast oriented, and that a regional show of books produced on the West Coast would be in order. Thus, in 1938 the Rounce & Coffin Club sponsored the Western Book Show, restricted to books printed west of the Mississippi River. It has continued to sponsor this show for over fifty years. Incidentally, the name of the club is derived from parts of the handpress. The coffin is the bed of the press on which the type is placed for printing, and the rounce is the crank that moves the coffin into place for the impression.

Zeitlin had been operating the Primavera Press as an adjunct to his bookshop for a couple of years when he decided he'd like to expand the publishing business and would need help. One evening when Phil Hanna and I were having dinner with him, he proposed that we join forces in an expanded Primavera Press. Phil Townsend Hanna was the respected editor of *Westways,* the Automobile Club of Southern California magazine, and an acknowledged authority on Western history. Jake asked him if he'd consider being the editor of the Primavera Press, while he would handle the distribution and publicity and I would take care of the design and production of the books. We all agreed, and the new Primavera Press was incorporated, with Zeitlin having 40 percent ownership, and Hanna and I each 30 percent. Our capital investment at the time was seventy-five dollars each.

The publishing program of the press was to be primarily devoted to Western Americana. At that time Zeitlin was committed to publishing only one book, a translation by Marguerite Eyer Wilbur of Alexandre Dumas' *A Gil Blas in California.* While Dumas had never been to California and knew little about it, his literary factory in France was known to accept manuscripts submitted by others and publish them under the name of Dumas as the author. Who the actual author was who had these adventures in California we will probably never know. Jake had already made arrangements with Saul Marks to print and bind 525 copies. Grant Dahlstrom was Saul's silent partner at that time, and the two of them had worked out a design without the western feeling Jake wanted. He then asked me to redesign the book, with Saul still to print it. I involved Paul Landacre to do a large map of the goldfields and a symbolic vignette for the opening of each chapter. He engraved these on boxwood, together with an adaptation of the first printer's mark used in the Americas in the sixteenth century by Antonio Espinosa. It was an anchor on which was superimposed the skull of a bull. I later adopted it as my own pressmark.

Saul, as a printer, was a perfectionist with little regard for time and, at that time, little experience in printing books. He had signed an agreement with Jake on Valentine's Day in 1933 to deliver bound books in two and a half months. He bought a secondhand Laureate press from Meyer's Showprint Company. It was the envy of all of us when it was moved into his shop. But it had been badly mistreated during the previous years, and the platen was sprung, making it almost impossible to get a good printing impression. Somehow, through ingenuity and perseverance, Saul was able to finish the book, but it took the rest of the year. He couldn't manage Landacre's large map of the gold regions, however, and I had to print it as a double fold tip-in frontispiece for the book. With the Christmas season approaching, Jake wanted books, and Saul promised that he'd deliver before Christmas. And he did, bringing a single copy to Jake's home late at night on Christmas Eve. While Jake was annoyed at the time, all was forgiven when the book was selected as one of the Fifty Books of the Year.

I was fortunate in having another book chosen for the Fifty Books that year. It was a small book of poetry I had written under the pseudonym of Peter Lum Quince. I was still reluctant to expose myself as a writer of poetry. I had used several other pseudonyms previously — Davie Dicker, James Beattie Pitwood, Betsy Ann Bristol, and Peter Mallory. For this book, *Fifteen Poems for the Heath Broom,* I had originally chosen the name Peter Quince from Shakespeare's *A Midsummer Night's Dream.* Paul Landacre had cut several engravings for decorations, and I found that my Peter Quince didn't aesthetically balance with the engraving he'd made, so I arbitrarily added the first three letters I picked out of the type case, which were Lum. For the same aesthetic reason, I shortened "hearth" to "heath." Subsequently, I wrote several other small books of verse as Peter Lum Quince, *The Year's at the Spring* (1938); *Fragments of Yesterday* (1941), printed by one of my students, Jane Frampton; *A Few More for the Powells and the Hea(r)th Broom* (1949); *A Summer Sequence* (1950); *Quince, etc.* (1976); *Poems for Marka* (1981); *Landacre & Quince* (1989), and *A Late Offering of Quince* (1990). The Santa Susana Press of California State University at Northridge published a selected group of poems in 1977 called *A Bowl*

84

XV POEMS for
THE HEATH BROOM

by PETER LUM QUINCE

LOS ANGELES : Ward Ritchie
MCMXXXIV

of Quince. The author in this case used the name Ward Ritchie. It was printed by Grant Dahlstrom's Castle Press, with illustrations by Irving Block.

The Primavera Press managed to survive for several years during the Great Depression, chiefly because there was no overhead and no salaries except for a small stipend we gave to Larry Powell to keep him and his new wife from starving. We found that by distributing a few shares of stock and including them on the board, we made everyone who worked for us happy and eliminated expenses. Carey McWilliams was our attorney and handled arrangements with our authors. He was later to become the very distinguished editor of *The Nation.* Cornelis Groenewegen, a boyhood friend, was our accountant. He worked for the Arthur Anderson accounting firm. Even with his doctor's degree, Larry hadn't been able to find a teaching job. He became our shipping clerk and a part-time employee in Zeitlin's bookshop. Eventually he went on to Library School at Berkeley and became librarian and dean of the Library School at UCLA.

This crew, along with Jake, Phil, and me, would gather often of an evening to discuss strategy, new manuscripts, and to enjoy one another's company. There was a convenient bar next door to Jake's shop where a hefty martini could be had for twenty-five cents. We'd precede each meeting with a gathering there. It was often a wonder that anything was accomplished, but the press managed to publish a number of worthwhile books before its demise in 1936. It might even have managed to weather the Depression except for one unfortunate involvement. The

Hollywood Festival Association was planning a spectacular presentation of *Everyman* in the Hollywood Bowl, and it authorized us to print 5,000 copies of the book, which they expected to sell at the Bowl during the summer months when the pageant was supposed to play. It lasted hardly a week, and we were left with almost 5,000 unpaid-for books. Groenewegen, at our next meeting, reported, "It is obvious that the sales for the past year and the current asset status are both very unfavorable. Unless this is altered, our current publication will probably be our last." *Everyman* was, and McWilliams dissolved the corporation in 1936. The residue of the books, including 5,000 *Everymans*, were turned over to me in lieu of my printing bill, and the Ward Ritchie Press became a publishing as well as a printing company.

In November or December of 1931 Hugh King arrived in town, hoping to start a handmade paper mill. His brother-in-law, Harry Beach, together with Hugh's brother John and Dard Hunter, had previously run the Dard Hunter Mill in Lime Rock, Connecticut. Private presses, printing precious books on handmade paper, had flourished during the fabulous twenties. Dard Hunter had done several books in the grand manner on historical papermaking, which he had written and designed, and for which he cast the type and made the paper, then hand-printed and bound. Hugh wrote me about the Lime Rock Mill where he had worked: *Granted our plans for the mill were ambitious, it is a tragedy we weren't allowed to bring the magnificent idea to a happy and successful conclusion. Had it not been for the economic chaos of that year, Lime Rock would have survived to become the shrine we dreamed of for those interested in fine paper and printing. We had plans to establish a printery as part of the project, with Hunter in charge of printing on his own paper with his own type. But we had no sooner gotten under way in 1929 when the market took its historic nose dive and about 95 percent of our market dried up. The figures as I remember them were that four hundred-plus books and pamphlets were printed on handmade paper in 1928–29 compared with less than forty for the 1929–30 period. We were competing with English, French, German, Italian, and Spanish mills, all long established and who could deliver the product here in the United States cheaper than we could produce it. It seemed hopeless, and we folded our tents and went our separate ways. Mine was to California.*

I first met Hugh in January 1932 when he gave a demonstration of handmade papermaking at the Los Angeles County Museum in Exposition Park. Soon after, the rest of his family moved west, including his older brother, John, and his two sisters. Mary subsequently married an artist who worked for me and Janet married my pressman, so I became quite well acquainted with the King family. Hugh made his first call on me, hoping to sell me some of the thirty reams of Dard Hunter paper he'd brought with him to California. This had been given him in lieu of back pay when the Lime Rock Mill had been closed down. When he first glimpsed my small shop, with hardly enough equipment to print a book, his expectations must have vanished, but he was asking for his beautiful paper not much more than what I was paying for ordinary machine-made paper. I bought quite a bit from him and printed several interesting books on this paper.

The first was Ronsard's *Sonnets to Helen,* translated by William van Wyck, whom I'd met in Paris. I also printed the Doves catalogue for Mrs. Millard and a book for the wealthy collector Estelle Doheny, *A Pre-Raphaelite Aeneid of Virgil in the Collection of Mrs. Edward Laurence Doheny of Los Angeles, Being an Essay in Honor of the William Morris Centenary 1934.* Mrs. Doheny had bought this manuscript through Mrs. Millard. It was truly magnificent, written in Morris' calligraphy with hand-painted illustrations by Edward Burne-Jones, who remarked that it was "to be so wonderful as to put an end to printing." It was indeed wonderful, and when later sold at the disbursement of Mrs. Doheny's books, it brought the princely amount of $1,320,000. My little book, along with forty-seven cards and letters written in appreciation of Mrs. Doheny's gift of the book, fetched $28,500. I also used Dard Hunter paper for the large folio, *To His Britannic Majesty King George V.* It was a single copy, printed as a gift to His Majesty on the twenty-fifth anniversary of his accession to the throne. It was signed by hundreds of former British subjects working in Hollywood, including Chaplin, Pickford, and Ronald Coleman.

The dream of the King brothers of establishing a handmade paper mill on the West Coast came to naught, and gradually their supply of paper was exhausted. Hoping to get more, I went to the source, Dard Hunter himself in Chillicothe, Ohio, sending some of the books I had printed on his paper. He wrote: *The three books were received today, and I am indeed grateful to you for your kindness in sending them. While the Connecticut mill will be abandoned, I will always be able to supply genuine Dard Hunter handmade paper as I am arranging to have paper made elsewhere. I do not believe it would be practical to make paper by hand in the present New England mill as there are many obstacles that cannot be overcome without constant personal supervision and I cannot devote all of my time to technical papermaking. I am now arranging to have my paper fabricated by a mill where my instructions will be definitely carried out, which will assure uniformity in thickness, tone, sizing, etc, etc.* Later on he wrote me: *While I have a large quantity of handmade paper here, I doubt if any of it would be suitable for your use. It all bears my own watermark and is in special sizes that I have used in the past for book printing.* He went on to suggest that he could arrange to have special paper made for me with my own watermark, but the minimum quantity they'd have to make and the cost was beyond my means at that time.

I am still uncertain how I first came to the attention of Estelle Doheny. It was probably through Dawson's Book Shop, her primary source of books during the early days of her collecting. Her husband, the wealthy oilman Edward Laurence Doheny, was an invalid at the time, and she had become interested in book collecting as a diversion. She had had a few small things printed by John Henry Nash, the flamboyant San Francisco printer. Most of his wealthy clients were no longer commissioning the expensive books he printed, and he was suffering. He considered Mrs. Doheny as a possible saviour. He printed an elaborate, bound prospectus of a projected printing of the Vulgate Bible, which he thought might appeal to her as a devout Catholic. He assured her of immortality if she would sponsor its publication by having her name coupled with his. He flattered her but made the greedy mistake of asking for too much too soon. He wanted a million dollar advance to get his project under way. She refused to become involved and immediately looked for another printer to do a forthcoming catalogue of an exhibition of her books to be shown at the dedication of the Edward Laurence Doheny Jr. Memorial Library. The family had given this building to the University of Southern California in memory of Mr. Doheny's late son, who had been a student there. The exhibit was to be held in the Treasure Room, a separate wood-paneled room in the library that Mrs. Doheny envisioned as her own and a place for her collection. It was so attractive that President Rufus von Kleinschmidt decided he'd like it for his own office and moved in. Mrs. Doheny never forgave him. Ironically, the great library she eventually accumulated and bequeathed, along with a handsome library building, to the Catholic Seminary at Camarillo might have gone to USC and not have been dispersed, as it later was.

I was selected to print the exhibition catalogue and continued over the years to do Mrs. Doheny's printing. The Doheny house was in an older part of town near West Adams, which had once been the elegant part of Los Angeles. It was a large Victorian house set in parklike surroundings, with a tremendous greenhouse filled with tropical plants and, at one time, even a small zoo. Mrs. Doheny was a remarkable woman, a busy executive who kept two secretaries occupied with her charitable activities, her business, and book collecting. Mr. Doheny was enfeebled at that time and was to die soon after. I recall having breakfast in the garden with them and seeing the great care she took in spoonfeeding him his breakfast. Later, as her collecting taste improved and she began accumulating books of great distinction from Mrs. Millard and the Philadelphia bookseller A.S.W. Rosenbach, she held many gatherings in the Crystal Room in the house to show newly acquired treasures. In 1940 we printed the first volume of the catalogue of the Estelle Doheny collection, which included about 4,000 books and manuscripts. When we printed the third volume in 1955, the collection had grown to over 7,000 volumes. As she grew older her eyesight commenced to fail. She was unhappy that she could no longer read her favorite prayers from her prayer book. We set and printed them for her in a larger-size type, which was satisfactory for awhile. But as her eyes grew worse we had to repeat this, each time in type a size larger.

My father died in 1929 and my mother in 1933, after which our family home was sold. I moved my printing equipment to an old ranch house in Los Angeles that my brothers and I had bought with the little money left in my mother's estate. It was a semi-historic building, once the foreman's house on the extensive cattle ranch that covered the hills of east Hollywood. It was perched on a hillside looking down onto a stream and a small mill. The stream had been covered and now two major highways wound down the little valley. The old ranch house and the mill were the only remnants left of the ranch, which had subsequently been sold and subdivided by Charles Canfield and his heirs. Canfield had been a partner of Doheny

Ritchie's Roadhouse

when they dug the first oil well in Los Angeles. His daughter, married to the movie star Antonio Moreno, still lived in the huge hilltop mansion that dominated the whole area now known as Moreno Highlands. In the early days of Hollywood movies, when Jesse Lasky needed a cowboy for one of his films he chose the foreman of this ranch, a fellow named William S. Hart, who became one of early Hollywood's biggest stars. With my brother's help we transformed the lower level of Hart's old house into an attractive studio, sometimes copied for movie sets.

It was a strange setup for a printing shop. Across the front, looking down into the valley, was a bank of large windows. In one alcove were two presses. The Washington handpress, which I used as a proof press, stood in front of one of the windows. Beside it was an old spinning wheel. Along the whitewashed walls was a series of type cabinets. Prints by Holbein, Eric Gill, and Paul Landacre decorated the walls, except for the one wall I kept free to pin up my own proofs, as I was designing after the manner of Francis Meynell. A grandfather clock's Westminster chimes kept us aware of time. Comfortable chairs and a couch looked upon the whitewashed brick fireplace. There was a rosewood grand piano and, on the floor, an oriental rug. In another room, up a few stairs, was my office and library. It became a hangout for a varied group of salesmen, artists, and loafers. Larry Powell aptly dubbed it "Ritchie's Roadhouse." John Cage used the piano to practice. After returning from Europe, he lived in a small house down the street and would appear almost daily to create and practice amidst all of our noise and bustle. With the presses clanging, the compositors pounding forms, the clock ticking, and the fire crackling, he would be creating a melody. Some of this experience may have influenced his later compositions.

Gloria Stuart bought the little mill house across the valley as a sculpture studio for her husband, Gordon Newell. When I had visited them a couple of years earlier on my return from Europe, she had just finished appearing in *The Sea Gull* in the Carmel Theater and had received an invitation to repeat her performance at the Pasadena Playhouse. She drove south with me, and a few nights later I was watching her performance, along with several motion picture scouts. She was immediately signed to a seven-year contract, starting a career of some forty films before she retired. I was happy to have my old friend so close by, together with a cluster of other artists in the neighborhood. The Walt Disney Studio was located only a few blocks up the street from us, and the Disney crowd enjoyed the ambience of the "Roadhouse," gathering there after work for a beer and conversation. It was inevitable, seeing all of the printing going on, that they would want to get involved. We conceived a Disney magazine as as a new outlet for fun and creation. A half-dozen of the writers submitted articles, several artists did caricatures and Fred Moore did some cute girlie drawings. We hand-colored the latter, setting up an assembly line, with six artists each applying a different color. We called our magazine *The Mousetrap* and it was such an instant success, we began making plans for succeeding issues. But we hadn't consulted Walt Disney, and we hadn't considered what his

95

Disney

T H E M O U S E C R A P

reaction might be to a slight jest we had played upon him. For the copy we sent him we had printed a new cover, with the title changed to *The Mousecrap.* It evidently didn't set too well with him, and there were no more *Mousetraps* nor *Mousecraps.*

In 1924, when I was a freshman at Occidental College, Gordon Newell and I were invited to a meeting of a literary society at Pomona College. It was a special event, as the group had been invited to the home of the authors, Paul Jordan-Smith and his wife, Sarah Bixby Smith. Mr. Smith entertained us with his sparkling wit as he talked about James Branch Cabell and his controversial novel *Jurgen,* which had been banned in several states. It was several years later, after I had become interested in southern California history, that I read Mrs. Smith's *Adobe Days.* It was the charming story of her girlhood in nineteenth century southern California. The Bixby clan had herded sheep to California in the 1850s, and had stayed on to acquire huge hunks of land, including the ranchos Los Cerritos and Los Alamitos, where Long Beach is now located. I had not seen her since that evening in her home in Claremont, until at a party given by Jake Zeitlin in his rustic place in Echo Park. We had a chance to talk and enjoyed each other. She was genial, outgoing, intelligent, and interested in the people and events going on in our town. She and her husband had left Claremont and now lived on Los Feliz Boulevard, not far from my place. Jake had picked her up and brought her to the party, and in the course of the evening asked me if I'd mind driving her home.

It was my pleasure, and on the way she mentioned a young daughter who was just recovering from an unhappy college romance and youthful marriage. She suggested that I might like to meet her sometime. I was not particularly interested, but not long afterward Jake invited me to another party, telling me that Mrs. Smith had asked that I be requested to bring her daughter. In this way I met Janet. We were later married and had two children, Jonathan and Duncan, and lived together happily until she divorced me twelve years later. It was an amicable divorce, and we continued on the most friendly terms until we both remarried and she moved east. Our life together was always interesting, but they were lean times during those Depression years for a girl who had been brought up in comparative luxury. After our marriage I built an additional family wing onto the old ranch house, with home and business continuing to be intertwined. In 1936 the press was moved to a two-story building a block away. Gone was the former easygoing charm — we became more commercial.

Gregg Anderson, who had been so very helpful in giving me guidance when I was first getting interested in printing, had departed the Huntington Library to enter Pomona College. He stayed for only a year and a half before leaving to work for the Grabhorn Press in San Francisco, where he set type for its great *Leaves of Grass.* Following this experience he went east to work with Harold Hugo at the Meriden Gravure Company. When his wife contracted tuberculosis, he returned west and started working for me. When it became obvious we needed a more efficient press, he contributed $500 toward the purchase of a Miehle Vertical press, and I made him an equal partner. We subsequently called the commercial printing division Anderson & Ritchie, retaining The Ward Ritchie Press name for our publishing. While still working at the Grabhorn Press he had written me: *Just what do you think of fine printing by now? The last six months have seen me gradually losing my awe and reverence for most of it. I have been reading De Vinne and I have become a De Vinne fan; there was a man who knew more about the making of books than all of the modern practitioners thrown together. Scholarship, learning, intelligence, those are the things that now seem to me the ingredients of a good book: more than all the art in the world they are the essentials. I find Ed's utter disregard for the logical arrangement and correctness of the text getting on my nerves now and again; his inability to make a book that can be read standing in the way of my appreciation of his other good and true qualifications as a bookprinter, so I busy myself at learning how to spell, tracing growth of the footnote in an effort to imbibe a learning that will make me competent to print McKerrow's next opus — have you read* An Introduction to Bibliography*? — or a book of poetry.* He contributed his practical interest to the commercial printing, while I devoted myself to the design of the books and association with authors and artists.

In the spring of 1934 I was informed by Robert Schad of the Huntington Library that I had been elected to membership in the Zamorano Club. The first meeting I attended was a special one held at the director's home on the Huntington estate. Max Farrand had succeeded George Watson Cole as the director of the library and had invited members of the Club to meet at his home to view his extensive collection of the etchings of Armand Coussens. A few years later, on the occasion of a visit to southern California by the eminent Boston printer Daniel Berkeley Updike, Farrand again invited the Club members to enjoy his hospitality and view his collection of Updike's Merrymount Press books. Farrand asked me to be the speaker for the occasion, prescribing that the subject of my talk be on "printer's flowers." It was of some concern to me, being asked to speak before Updike, a man who was an acknowledged authority on the subject.

Fortunately, Farrand had given me several months' advance notice in which to prepare. I was thus able to read not only what Updike had written on the subject but also articles by Francis Meynell, Stanley Morison, Douglas McMurtrie, and others. I gathered numerous examples of the use by printers of these flowers as decorations in books and other printed pieces. From my own shop I showed how Alvin Lustig had been able to combine geometric units into striking abstract art forms. There were also the giant pictures that Albert Schiller composed, sometimes taking as long as a year to complete. Bruce Rogers, in his later playful years, enjoyed puttering with these flowers and made many whimsical arrangements, combining them with dingbats of toys and animals.

A variation of the traditional printer's flowers was devised by William Addison Dwiggins. I have admired his work as a book designer as the most innovative and original of any this century. He added a new dimension to decoration by creating a series of small units cut in stencils, which he could combine into pictures and decora-

100

Elements for making pattern to be used with type
Cut in celluloid — .075 in mostly
These discharges show the stencils as cut
Some of them are made with the idea of touching in the tie-breaks with a pen, to finish — but mostly used as cut
WAD August 26 1946
The property of Ward Ritchie — will he consent to having it copied via photostat, if the occasion for so doing might arise? WAD
repeats
repeats
stems — bases — buds
repeats
repeating bands
Geometrical Set

tions as others had done with type units. When I wrote him of my predicament in having to give this talk with Updike in the audience, he was extremely helpful. He not only sent me many examples of his use of the stencil units, but went to the time-consuming task of stenciling the many hundreds of elements he had created and cut in celluloid. This display, more than four feet long and beautifully arranged, became the focal point of my talk and allowed me to introduce some new material on the subject. He wrote me in explaining the process:

The printer's ornaments game started years ago, inspired by the sprigged designs on India cotton prints. The first attack was with little woodblocks of buds, leaves, etc., like small rubber stamps. These were too hard to manage in Higgins ink on paper. Another attack later, via small stencils cut in celluloid, was easier. I cut a lot of elements and got pleasant results. You could see through the celluloid how to combine them. Another later dodge was to see what you could get by combining pure geometrical elements, circular arcs, angular bits, triangles, etc. — these in celluloid also. Cut a lot of them. Carl Rollins calls the results my concrete-mixer art. The first good word on these came from D. B. Updike who spoke nicely about a set of ornaments in a book: Paraphs *by H. Puterschein. He said they were twentieth century Pillements — which was nice because I was keen about Pillement. Paul Beaujon (Beatrice Warde) had an article about them in* Fleuron, *No. IV. Used them a lot — still do.*

Years later I used a quotation from Dwiggins to finish a Commonplace Book I was printing on my handpress. It read, "It was a grand adventure. I am content." Dorothy Abbe, who was Dwiggins' helper in his later years, wrote to tell me that it was also carved on his tombstone.

The Zamorano Club had been formed by a small group of Los Angeles bookmen as a generous way of helping a proud and impoverished legendary bibliophile to live comfortably. W. Irving Way had been well-known as a publisher of the firm of Way and Williams in Chicago during the waning years of the last century. He had been a charter member of the Grolier Club of New York and a founding member of both the Caxton Club and the Duodecimos in Chicago. He had come to California in 1904 and during the following years had made a meagre living selling subscription books and as a book scout. He was getting old and found it difficult both to eat and pay the six dollars a week for his room in a dilapidated small hotel on Figueroa Street. He subsisted mainly by selling book treasures accumulated in earlier years. He would take a book at a time to Dawson's or Jake Zeitlin's bookshop and get enough to get by for another few days. On one of these occasions, Gaylord Beaman, an avid collector of a great variety of books, noticed Way selling a precious volume to "Father Dawson." Beaman became interested and before long was taking Way with him to have lunch with two or three other bookmen at the old Alexandria Hotel on Spring Street. The men were intrigued by the tales Way told about books and bookmen he had known. Wondering how they could help this charming old man without offending his dignity, they conceived a book club fashioned after the Grolier Club of which Way had told them. They gathered a few similarly minded men and formed the Zamorano Club, named appropriately after the man who introduced printing in California, Agustin

V. Zamorano. They appointed Way to be librarian of their nonexistent but proposed library, with a monthly salary sufficient to meet his needs. He was much cherished by the members until his death two years later.

The Public Works of Art Project started in 1933 was a great stimulus to art in this region as well as the rest of the United States. Mr. Edward Bruce had been able to convince President Roosevelt of the need not only to help artists financially during the Depression, but also to give impetus to the recognition of art and artists. The southern California area was administered by Merle Armitage as regional director, with Dalzell Hatfield, Arthur Millier, Louis Danz, Merrill Gage, and Millard Sheets on his executive committee. Between 150 and 200 artists were given projects and paid on a weekly basis. It was understood that "the eligibility of these men and women has been determined entirely by the quality of their work rather than by financial need in accordance with the policy laid down by Washington headquarters." Yet it helped a great many otherwise semi-impoverished artists to keep working, and it brought art to a great many public buildings, schools, libraries, post offices, etc. Much of this work has now disappeared without much trace or record. At the time it brought artists together in an unusual camaraderie.

I did several pieces for the project, including the Declaration of Independence, a Lincoln speech, and a tribute to Edward Bruce. It also brought me into association with many artists who eventually illustrated books for me. Paul Landacre, of course, had already done wood engravings for several of my books, but for him this additional government aid was a true godsend. He had been an outstanding track athlete at Ohio State University when he was struck by a debilitating disease that left him permanently crippled. He began to draw as therapy, then found he had a talent and gradually expanded his skills to include wood engraving, which he taught himself since there were no teachers in southern California where he then lived. His technique was awkward compared with that of the engravers I had watched in Schmied's *atelier* while I was in France, but it worked for him. He became one of the finest engravers in America and a wonderful artist to work with. I could explain to him what I wanted and it was always better than I expected.

In March of 1934 the Los Angeles Museum held an exhibit of the work of the WPA artists, showing 214 pieces ranging from paintings, sculpture, printing, and prints to photographs by Edward Weston. The Polish sculptor Stanislaw Szukalski did a heraldic American eagle design for the cover of the catalogue. It appeared to me as a possible decoration for the title page of Laura Riding's *Americans,* which I was then doing for the Primavera Press. With his permission, I had a talented young fellow by the unusual name of E. Digges Graves who was working for me engrave it on wood, and I printed it in a striking red, which made a brilliant title page.

Szukalski's life was one of self-destruction. In youth he was acclaimed and famous. In age he was lost and forgotten. He was brilliant, self-centered, and as egotistical as any person I have ever known. He had great natural ability as an artist, and when he came from Poland to Chicago at the age of only eighteen, he made quite a stir in art circles with his bizarre drawings and sculptures. He married a wealthy socialite and had two sumptuous volumes of his work published — one by Covici-McGee in 1923 and a subsequent volume by the University of Chicago Press. He had come to California in the early 1930s and lived with his wife in a mansion in the Hollywood hills, with plenty of space for the enormous bronze pieces he had created. When Gordon Newell and I first visited him, I was overwhelmed, especially by a mammoth statue of a crouching Mussolini with his square jaw extended and an arm outstretched. He soon abandoned his wife and the luxury she afforded him for another woman. He stored his sculptures for awhile and then, deciding to return to his native Poland, he shipped them there. When the Nazis invaded Poland, Szukalski, his wife, and daughter barely escaped to return to the United States, leaving his life's work, which was probably melted down for munitions by the Germans. In addition to the American eagle, he did an exotic alphabet for me that Digges Graves also engraved on wood. It was so very strange that it wasn't until fifty years later that I found a way to use it. He was great at portraiture and did portraits of Larry Powell, Al Fisher, and of Janet and me. These portraits are now at the Clark Library.

David Alfaro Siqueiros was another artist from a foreign country whom I came to know. He was from Mexico and had been preceded here by several other countrymen, among them José Clemente Orozco, who had done a fresco of Prometheus on the ceiling of the dining room at Pomona College; Alfredo Ramos Martinez, who did an unfinished mural at Scripps College; and Diego Rivera, who did some work in San Francisco. Siqueiros was a belligerent Marxist and was almost always in trouble in Mexico, spending much time in prison. After being released from a term in 1932, he came to Los Angeles with his pretty wife of the moment, Blanca Luz Blum. While in prison he had spent some of the weary hours cutting woodblocks with a penknife on what appeared to be the ends of apple boxes. He managed to sneak out a dozen of these blocks when he was released and brought them along with him when he came to Los Angeles. I printed several sets of proofs for him on my Washington handpress, using some of the Aurelian paper I had left over from *The Colophon* job. Jake Zeitlin sold the sets for a nominal sum.

Siqueiros had developed a new method of doing murals, using plastic materials and spray guns. He generously shared his discoveries with the local artists who worked with him on several murals. Millard Sheets, Fletcher Martin, Barse Miller, and Phil Paradise were among them. A blank wall always attracts a muralist or even a graffiti artist. Siqueiros spotted one down on Olvera, the historic Mexican street in old-town Los Angeles. It was the second-story side of one of the old buildings, and he arranged to do a mural on it with his coterie of eager followers. He designed it depicting a naked Indian body crucified on a double cross with an eagle perched atop. I climbed the ladder to the roof of the

107

David Alfaro Siqueiros

Courtesy El Pueblo de Los Angeles Historic Monument

adjoining building several times with Fletcher Martin. I was only an onlooker, but I was fascinated watching a half-dozen men working together under Siqueiros' direction. After completion it didn't last long. It was too radical for its time, and the Olvera Street Association had it painted over. Now, sixty years later, it is being reconstituted.

Remsen Bird, the president of Occidental College, was a frequent visitor and a constant catalyst. Noting the unusual number of artists who congregated around my place, he suggested we form a club of artists. It seemed a good idea and we proceeded. The old ranch house was getting a bit small for the press, and when a two-story building a block away on Hyperion Avenue was offered for $6,000 with only $600 down, I bought it and moved the press there. The empty studio seemed made for a gathering place for the club. Gordon Newell from across the road; Archie Garner, another sculptor; and several lads from the Disney Studio helped build a long table, which we set along the windows looking out over our small valley.

A year or so earlier a colored gentleman from the Deep South had climbed the hill and knocked at the studio door, asking if I might be Colonel Jackson. When I answered, "No," he said, "I'spose to be working for him, but I've walked all the way up and down this street and I can't find him. I'se tired and hungry. If I picked a few weeds for you, could you give me a bite to eat?" I agreed. At dusk when I paid him a pittance for his work and fed him, he pathetically said he had nowhere to sleep that night, and noticing a small storeroom under the house, he wondered if I'd mind if he slept there. And thus J.D. Hicks became a part of the family for the next few years.

We had grandiose plans for "The Club," as we named it. We'd have lunch every day prepared by Chef J.D. Hicks, just as the downtown University and California Club had. The boys from Disney would join with the local artists in a jovial gathering each day. And so it started. But that phase didn't last too long. For one thing, J.D.'s menus were not very appetizing, and meanwhile the Disney Studio began its move to Burbank. We abandoned the luncheon idea and substituted a Thursday night sketch class, which proved to be a most happy substitution. We'd get a model and a case of beer. We'd sketch for an hour or two, send the model home, and then sit around telling tall tales until the beer was gone, usually about one o'clock in the morning. Delmer Daves was a great storyteller. He was also the most affluent of the group and would pay the modest two dollars a month dues for those who were at the moment a bit hard up. These dues paid for the model and the beer. When we had a special party, which was often, there would be another dollar assessment. Delmer had been at Stanford when I was there during my sophomore and junior years. He was an all-purpose man on the campus — writing, making posters, acting, teaching, creating the rallies, and about everything else. After graduation he came to Hollywood doing bit acting. Soon he was writing and finally directing and producing many pictures. He had a great and generous heart. He took particular care of Paul Landacre, sending a carpenter to keep the house from tottering on a hillside, helping with the neighbors to reroof it. He, along with Jake Zeitlin, induced a number of persons to pledge to buy a Landacre print once a month, which finally gave Paul freedom from financial worry. Years later, when Paul's wife Margaret was dying of cancer, Delmer called me saying he wanted to do something to help. He was directing and producing *Youngblood Hawke,* a picture about a writer, based somewhat on the life story of Thomas Wolfe. He arranged for me to do a series of dust jackets for the author's various books, each using a Landacre print, for which Delmer paid Paul enough to cover the hospital costs.

Fletcher Martin

Fletcher Martin was also a consummate storyteller, and he regaled us with stories of his boyhood, growing up on a farm in Idaho and working in the printshop of the various small-town newspapers his father would buy and run for awhile. Whenever anything went wrong, his father would give him a thrashing, since he was the oldest boy. Fletch finally ran away, hoboing for a year before joining the Navy. He was only seventeen but lied in order to join. He was not only the light heavyweight champion of the Pacific fleet but discovered he could amuse his mates by drawing suggestive pictures for them. He came to Hollywood after his stint in the Navy and worked in a printing shop that did a lot of movie work. He started painting scenes he remembered of his boyhood and Navy days, developing his native talent without the help of any formal instruction.

The Club was a handy reservoir of artists for me when I needed an illustration for a book or an announcement.

Fletch and Denny Winters

Leigh Harline

Ward and Herman Cherry

Janet

Alex Brook and friend

Barse

Martin did some drawings for the book *Of Una Jeffers* and for *John Gutenberg,* a book I wrote to commemorate the five hundredth anniversary of the advent of printing. Paul Landacre was ever available and illustrated a dozen or so books with his superb engravings. Barse Miller did an illustration for *The Gutenberg Address* by Christopher Morley. He, like Fletcher Martin, was a quick-sketch artist. Perfect drawings seemed to come from his brush or pencil without effort, study, or need for revision. The Book Club of California had asked me to print for its members a group of early California stories by Bret Harte, Joaquin Miller, Ambrose Bierce, and several others, to be called *Ballads of Eldorado.* I had a crazy idea of a continuous illustration that would run from the back cover around the spine to the front cover, and then continue onto the endpapers, the halftitle pages, until it ended on the title page. It was to be a panoramic introduction to the book. I made up a dummy book, hard-bound with blank pages, and took it along to the Thursday night sketch class, along with some early photographs of San Francisco. While the others were busy drawing the nude model, I dragged Barse to a corner and asked him if he'd like to illustrate the book following my unorthodox plan. He agreed to try. Unlike most of the members, he liked to work with watercolors rather than charcoal or pencil. He mixed some colors and, taking my dummy, started on the back cover with an early San Francisco street scene, which he continued around the front cover and onto the endsheets, where the illustration climbed Telegraph Hill and onto the next page, where it descended to the bay and onto the double title page, where it ended at the ferry building and a ferry boat plowing the bay. In no time Barse had completed the whole series and was ready to join the others painting the model. When he brought me the finished drawings the next week, they were almost identical to what he had so hastily and accurately sketched.

The war scattered many of the members and about the same time we moved to La Cañada, leaving the Club with no handy place to gather. Except for a few reunions, those happy associations faded into memories. The facility with which both Fletch and Barse were endowed served a good purpose during the years of the war. Barse was in the Navy as an official artist for the Pacific Fleet, and he made thousands of sketches of life aboard the ships and of battles and landings. Fletch was assigned by *Life* magazine to join the troops during the North African campaign, recording that action. Later he joined the Normandy invasion, sketching that perilous adventure. *Life* had given him a camera to record some of the sights and activity, but he found his sketches to be more satisfactory. Several of these were later developed into paintings, and *Life* used many in the magazine, including an eleven-page spread in the Christmas 1943 issue.

During the holiday season of 1937, Jake Zeitlin asked me to print a piece by D.H. Lawrence that Alvin Lustig had designed for him. It was an intriguing use of type and geometrical-type ornaments, ultramodern in concept. I visited the young artist in a room he'd rented as a studio in a dingy, rundown building on West Seventh Street. I climbed to the second floor and walked down a bleak, dark hall and was completely surprised to enter a cheery workroom. Lustig had painted the dreary walls a bright, fresh color and had created an attractive place. The furnishings were sparse but beautifully functional. He'd built for his work a semi-desk, which accommodated a handful of trays of type and the geometrical ornaments with which he developed his designs. He was a tall, slim, dark-haired lad with piercing eyes. We became immediate friends, and I was intrigued with the ingenious designs he made with a minimum of type material.

His artistry was his temporary undoing. When the landlord saw how attractive Lustig had made his quarters, he concluded that the rent was too low and doubled it. This infuriated Lustig. He unburdened his wrath to me, and I offered him a spot in my office rent free. As a boy in high school, he'd been an amateur magician, so successful and sought after that he hardly had time for classes. He later attended Los Angeles City College, where he took a course in printing from Richard Hoffman. He spent some time at Art Center School and three months at Taliesin with Frank Lloyd Wright. That interlude didn't work out. Wright wanted things done the "Wright" way, and Alvin insisted on doing them the "Lustig" way. Needing a pleasant place to work, he completely redesigned my office and had me wielding a paintbrush alongside him. He arranged a nook in one corner for his combination working desk and type cabinet. He paid no rent and received no salary. He had his own clients and would set up his distinctive modern layouts, which we would print for him. During the two years we shared the same quar-

116

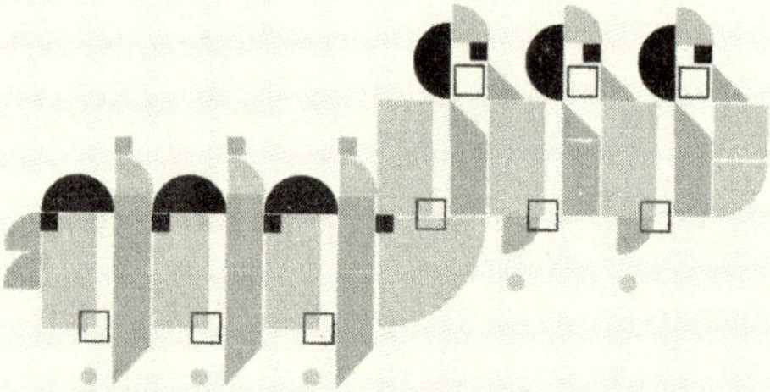

of a will to power and all of the mundane trivia, the whiles this veri-poet sits on his rock and gazes bare-eyed at the sun. Deep within the heart of him he knows that man is not worth the gunpowder to blast him from the earth, a thing that he is making a creditable attempt to do at present.

We may bawl all of the platitudes we like concerning the goodness of man, but one war, such as was the great war, discounts all of man's goodness to date. Man will spawn and create future generations to behave as he has always behaved. The small soul will blind itself to all of this and will continue to deceive itself in mouthings that have nothing to do with the words that spring from the heart of a veri-poet.

ters, we talked a lot. He would relate many of the intimate details of his life. While he was a Jew he had discovered Christ and was absorbed not by Christianity but by Christ's philosophy. At that time he was convinced that he was the possible Messiah here to save the world.

I had printed a couple of books by William van Wyck and he continued to send me manuscripts. He had left Paris and was living in San Francisco, where he was a literary commentator on one of the radio stations. One of the manuscripts he sent me was a talk he had given on Robinson Jeffers. I liked it but felt it was too short for a book. I puzzled over it for some time and finally found a possible solution. I asked Lustig if he would make up from his geometrical typographical material a running series of decorations to occupy the upper portion of each page, flowing as a symphony. He agreed, and we made a curious and delightful little book with his playful decorations printed in several colors. While Alvin mostly conducted his own business from his small corner stall, I'd occasionally prevail upon him to create some of his typographic designs for books I was working on. For *What Are Masterpieces* by Gertrude Stein, he made a simple decoration for the title page. For Aldous Huxley's *Words and Their Meanings,* he did more elaborate designs for the title page and also for the cover and the jacket. He did a large design for the cover of Herman Cherry's *Scrapbook of Art* and for me a large print for framing. The last book he illustrated before he left to have his own studio in the Brentwood area was *The Ghost in the Underblows* by Alfred Young Fisher.

There was much shuffling around and groping for stability among my friends during the middle years of the 1930s. Larry Powell, intent on teaching after he had received his doctor's degree in literature, found the only job he could manage to get upon returning to California was at Jake Zeitlin's bookshop. As his interest shifted more to books, he followed the suggestion of Althea Warren, the city librarian, and went to library school at Berkeley. Our friend from grammar school and high school days, Dr. Alonzo Cass, generously lent him the money. After receiving his library degree, he was able to find a job in the library at UCLA. It was a relatively small library in books then, though housed in an imposing Romanesque structure now named the Powell Library after him. He arranged to have one of my publications, Van Wyck's *Robinson Jeffers,* chosen for the ceremony accompanying the acquisition of their 300,000th book. After he became the head librarian, the number of books in the library soared to over two million and is now nearer seven million.

Al Fisher managed to get a job teaching English at Occidental College through our friendship with Dr. Stelter. Mary Frances became a discontented faculty wife, writing pieces for her amusement during faculty functions. Gloria Stuart and Gordon Newell were living in Hollywood, where Gloria was making pictures at breakneck speed, forty-one in seven years. Dillwyn, or Timmy, and Gigi Parrish were new and good friends. Gigi was in pictures along with Gloria. They were both what was called Wampus Baby Stars for publicity sake in their first years. Timmy was a writer and amateur painter. His sister was the novelist Anne Parrish and his cousin the painter Maxfield Parrish. He was homely but gentle and seemed to be attractive to beautiful women. He had been a tutor to Gigi in her New Hampshire home when she was only twelve. He was in love with her then and when she was sixteen eloped with her, the pair fleeing to California on motorcycles. After a few years in Hollywood, their marriage began to falter. Gigi was working most of the time and often on the road as a Wampus Star creating publicity. They had a place in Laurel Canyon, and when Timmy decided on a trial separation and left for the East to stay for a while with his sister, he asked the Fishers to stay with Gigi so she wouldn't be alone. It was summer vacation for Al, and he and Mary Frances were only too happy. Before long Gigi started divorce proceedings. Timmy, in despair, wired Mary Frances to accompany him and his mother on a trip to Europe, she to act as a companion to his mother. She accepted. While on the trip there began a love affair that did not end when they returned home. Timmy soon returned to Switzerland, bought or rented a château, and invited the Fishers to join him. Al left his job at Occidental and the three of them lived together in Switzerland. It was an arrangement of which novels are written, a beautiful woman living with a handsome husband and a homely lover in a storybook château nestled in an isolated valley among the rugged mountains of Switzerland. M.F.K. Fisher, in all of her autobiographical writings, never touched on her *ménage à trois.* It finally became unbearable for Fisher. He returned to the United States to accept a teaching position at Smith College in Massachusetts.

The lives of my many good friends were being rearranged *en masse.* Dillwyn Parrish and Mary Frances were married in Switzerland, where they continued living for a while. During World War I, while driving an ambulance in France, he had injured a hip. It later became gangrenous and his leg had to be amputated while he and Mary Frances were still living in Switzerland. They returned to California to an almost desert hideaway near Hemet. He painted and they wrote a novel together. Eventually, the pain was too great and she watched as he took his life. She adored him, but after his death she wrote Gigi, "You were his true love and he was mine."

Gigi married John Weld, a writer and former stunt man, who had assiduously pursued her even when she was still married to Parrish. Hollywood life and being the almost anonymous husband of a motion picture star was too much for Gordon. Gloria also found it annoying to return home exhausted at night after twelve hours on the set to find her husband entertaining friends. They were divorced. She was remarried to Arthur Sheekman, a writer who did most of the Marx Brothers films. Gordon, who had begun teaching in one of the local art schools, married one of his pupils, a White Russian from Manchuria named Amelia Bubeshka. Powell remained married, though barely at times, and I for the time being.

After fleeing from his unhappy sojourn in Switzerland, Fisher arrived at Smith College, an all-girls school in Northhampton, Massachusetts. He was young, inordinately handsome, charming, and a poet, almost a reincarnation of that other poet, Percy Bysshe Shelley. He soon surrounded himself with a coterie of adoring students. The late Ray Billington, a senior research fellow at the Huntington Library, had been on the Smith faculty at the time of Fisher's arrival. He told me, somewhat wistfully, how Fisher had become academically spoiled by all of this admiration. Each morning while he was having his breakfast on campus, a flock of girls would gather around, gasping with pleasure at every bite he took and squealing with delight at his every word. He had an image to maintain. He married two of his students. Although neither marriage lasted, he had a son by one and two daughters by the other.

Powell, Newell, and I had been in awe of Fisher's talents and of the lengthy epic poem he had written while in Dijon. We kept suggesting that he should allow it to be published. His answer was always an emphatic "no," his excuse being, "It needs more work, its length would frighten readers, it is only a fragment of what I had planned." When the first M.F.K. Fisher book, *Serve It Forth,* was published in 1937, he relented, possibly annoyed that his former wife was being published using some of the stories with which he had once regaled us. Soon after, on one September night in 1938, the three of us, Powell, Newell, and I, over a bottle of good red wine, agreed to see that it was published. Powell took on the job of editing and preparing the manuscript for publication. It had to be condensed considerably, else it would run to an impossible length. Even so, it made a book of 300 pages. I was to print a prospectus with the hope of getting both sponsors and pre-publication orders. Gordon Newell would do the addressing and mailing.

First we sent selections from the poem to several people whose judgment we respected, for comments and testimonials we could use in the prospectus. The replies were most encouraging. Robinson Jeffers wrote, "*The Ghost in the Underblows* is a remarkable poem and travels a wide orbit. It has exceptional qualities of imagination and irony, and distinguished expression." His wife, Una, wrote, "It is a remarkable poem." William Everson, also a poet, said, "You get the conception of an infinitely sensitive and intelligent man laying his ear to the earth and writing verbatim every delicate response and flux that twitches his being. The poem is sensitive, yet so big you have to take it in its books to get it. I cannot say whether Fisher has written an *Inferno* or a *Wasteland,* but he has written humanity, that much I know." Jens Nyholm of UCLA wrote, "The excerpt of *The Ghost in the Underblows* shown to me would indicate the arrival of a new American poet of unique stature."

We prepared an elaborate twelve-page sales presentation, and, since the book was beyond our limited means to finance, we urged subscribers to send payment for the book in advance at six dollars a copy. Since there were only 275 copies for sale, it would give us $1,650 to cover the production, printing, binding, and mailing costs. Today that would hardly pay for the prospectus, and it certainly left nothing for profit nor author's royalties. The orders came in slowly and we were quite discouraged until we got a phone call from Dr. Elmer Belt, a local book collector,

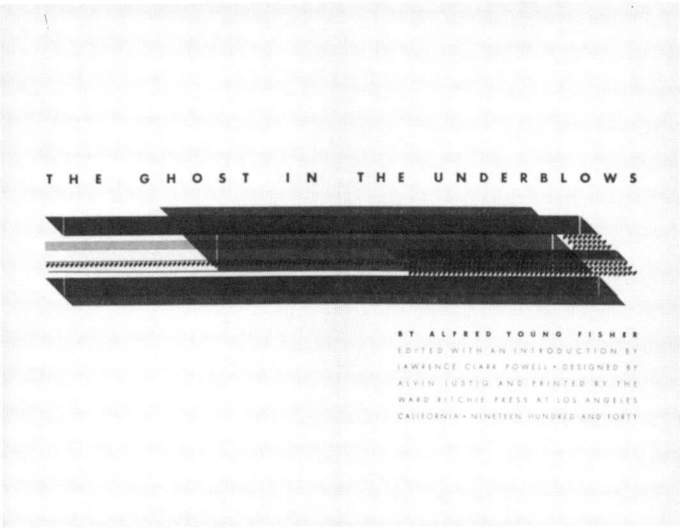

123

The Ghost in the Underblows

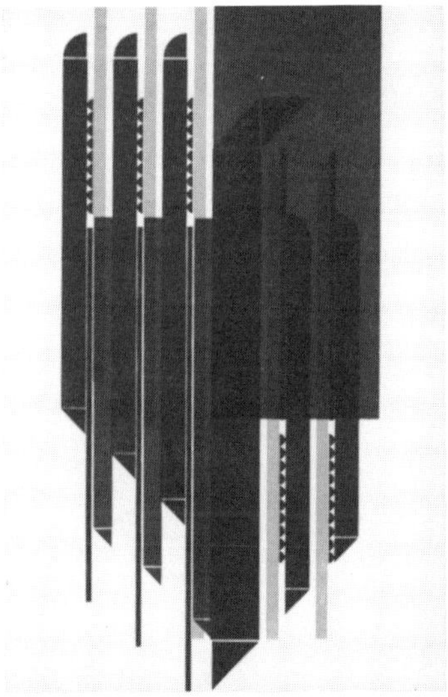

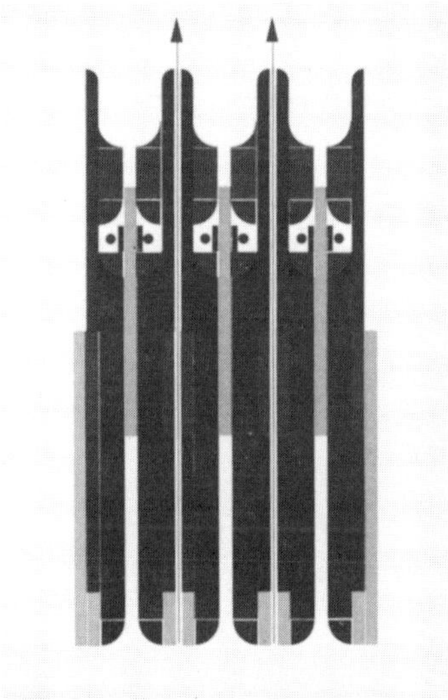

inquiring about the progress and agreeing to subscribe for all unsold copies. With that assurance, the production of the book was under way.

Alvin Lustig had created for me a striking typographical decoration for the cover of the prospectus, and as I was planning the book I decided I wanted him to illustrate it with a series of his monumental abstract creations. The text was divided into ten books. Fisher's original concept was to paraphrase the books of the Bible. Powell had condensed the text to the first ten books. As Alvin progressed with his ten heroic decorative pieces for each book, I realized that it was his book and let him continue to design the title page and binding. It was a wise decision on my part inasmuch as the result became a landmark in modern book design.

While the book received typographical acclaim, it went unnoticed in literary circles. The title was misleading. It might have been a mystery tale. The length of the poem was forbidding. I doubt if any but ourselves ever read it through. Possibly our judgment was tinged with friendship. It was a disappointment to Fisher. He watched his former wife's publishing triumphs while he never again wrote anything of consequence, although he privately printed a few pieces in which he had attempted to out-Joyce Joyce, and a few doggerel verses. Typical is the final stanza of a waggish one he wrote about the four of us who had become close friends and drinking companions during the 1930s. We had nicknames — Fisher was Shagbag, I was Richman, Newell was Monk, and Powell was Goose.

So, boys, let glasses cry,
Who drinks is dry?
Less than a tun is sacked hypocrisy!
Come, guttlecrack the snout
Less spirit be in drought,
My boys!
Bang the bottle, sound the horn,
Souse the two-eared unicorn,
Wash its head and slosh its shoes,
My Shagbag, Richman, Monk and Goose!

Not far from Northampton, where Smith College is located, is the small town of Cummington where Miss Katherine Frazier had her Cummington School devoted to music and the arts. Fisher had become acquainted with her and on an occasion when she mentioned she would like to include printing in her curriculum, he suggested she get in touch with me. I had a letter from him in 1939: *Some time ago I took the liberty, on the occasion when I heard she was hoping to begin at her school a printing project, of indirectly telling Miss K. Frazier about you. She is, I think, a kind of pioneer. Her work is difficult; it will be, perhaps, a long time before it is at all widely recognized. Yet, there are those who are in a position to know about her and Cummington School (of whom I am, perhaps, one) who think very highly of it and her. The point of this is that I should be pleased if you would take time — should she come to see you (on her first vacation in twenty years) — to give her all the help you can with her project for establishing a handpress.*

She came West and visited me in February 1940. I gave her as much advice as I could, suggesting a Washington handpress, which would be both ornamental and more injury-proof for neophytes than a mechanical press. I suggested she get some classical types such as Centaur or Bembo and listed what essential tools and material she would need. She wasted little time, writing me toward the end of April: *Our Washington Hand Press has been shipped to us and I am beginning to try and find some type.*

It was some time before I heard from her again. Then, in a letter dated July 20, 1940, came the following news: *The chief reason for my delay has been that I wanted to send you one of our first printing when I wrote. We have been having a good time trying to figure out everything. We have not had any skilled instruction on the use of the press yet, but did get a little advice from an ordinary printer. We have delayed going ahead because Paul Wescott of the Hill School has promised to come. But eventually our enthusiasm got the best of us, and we have struck off some of the enclosed little quotation. Harry Duncan has done practically everything about this.* She inquired further about obtaining additional Centaur type and a source of paper. She also enclosed their first attempt at printing, a quotation from *Ars Typographica:* "Simplicity is the last thing learned. It comes from simple thinking, not from a conscious attempt to be simple." It was printed on a Japanese paper with a large opening initial "S," which appears to have been cut on linoleum.

A month or so later I received a package and another letter from Miss Frazier: *I am sending you today a copy of our first printed work,* From This Hill, *and I hope you will both enjoy it and criticize it. I shall send you as good a copy as we made so that you will know at what point to begin. And I ought to explain a little. In the first place we had a sort of compromise situation; we needed first of all to fulfill our promise to our young contributors to publish their work, and so getting it out as literature took precedence over doing fine work as printers. (We certainly should have practised before attempting to actually print.) Then the students who could stay to work on the press had to leave at a specified time so that we had that limitation of time. Also we had a very limited supply of paper and could not practise as much as we should on the paper to be used; and we found that what was right for a galley proof was no use at all on the paper we used.*

We had no instruction except from my job printer on how to put the type together and how to clean it. All the people who had promised to come to instruct failed. We did this whole little book in two weeks. And we learned a lot every day, but never enough nor soon enough. The last day's work was much the best. We only had enough type to set four pages. So we had to distribute as soon as printed in order to set up the next bit. We had to plan the pages before it was printed at all; and some pages had to be too long because of our limitations. We had a terrible time on account of a slight slur, or double-impression; not until the last day did we discover that the platen just slightly touched the tympan once in a while and thus caused our trouble.

I don't think we found out, in all our work, how to handle the ink. On damp days we had to stop because the brayer would not pick it up. And, of course, the very best part of all is still to come, the using damp paper. For the sake of certain things, such as the effect of our block print, we wanted the paper we used, but it did not adapt itself well at all to being wet. And with all our greenness it seemed best to postpone the damp paper until we had learned some of the rest . . . As I believe the ink for the damp paper is quite a different proposition, anything we might have learned about this job ink for the dry paper would not be of moment in the future.

I am trying to learn something about paper from some men in the vicinity who I hope to see next week. But I need all the advice I can get. And I certainly need advice about ink. We hope to do some very small things yet this fall and want it to be much better than this. I shall be most grateful for your criticism.

From This Hill had most of the faults experienced by eager beginners. Harry Duncan, one of the literary contributors as well as one of the participating printers, later on wrote of how after the summer session of the school, a number had stayed on taking turns keeping the stove burning to keep warm while alternating at work with the composing stick, the brayer, and the press. They enjoyed the experience but were not too happy with the final result and eventually burned most of the edition. I wrote Miss Frazier an encouraging letter with some suggestions and sometime later received another packet with several more examples together with a letter: *It is a long time since your good letter came, with its lenient comments and good advice about our printing. We have profited by the advice and have worked hard this fall; under separate cover I am sending you examples of progress. The care of the brayers with kerosene instead of gasoline proved most important. We had a hard time finding any ink as hard as you suggested. But I finally made a pilgrimage to Hawthorne House in Windham, Connecticut, where Edmund Thompson gave me a lot of time and advice. From him I found I could get that hard ink only in California; he gave me some to try and I have since bought some of Goodman in Oakland. We think it will be right when our conditions are better. But we have worked under the most adverse circumstances this fall — in an unheated barn, only having a little round oil heater, slight protection from dampness and dirt. But I think we learned more from that than if things had been just right. It has probably been too cold to use the very hard ink. When printing damp, it pulled the paper apart, took the surface right off.*

Miss Frazier suffered from a lingering illness and was bedridden for the last two years of her life, dying in May 1944. The Cummington Press eventually became Harry Duncan's. He developed his skill in sessions working with Edmund Thompson and with Victor Hammer, and he took The Cummington Press along with him the years he taught and printed at the University of Iowa and the University of Nebraska. He, together with Victor Hammer in Kentucky, Lewis and Dorothy Allen of California, and one of his students from Iowa, K.K. Merker, have been doing the finest handpress printing in America. He learned his skill the hard way. Unlike most private press printers, Duncan prints only unpublished texts.

In May of 1933, when I was still struggling as a young printer, I received a letter from Frederic Goudy, the eminent type designer: *A note from Mr. Carroll Harris (San Francisco) received this morning suggested that you might be interested in securing the exclusive rights of my "Village Text," the redesigned & recut "Aries" which I offered some months ago to Mr. Grabhorn at a ridiculously low price as such things go.*

I have been holding off, as one of the large N.Y. publishers was considering it, but who finally decided that times are too hard to incur the expense for the comparatively limited use they could make of it — it is essentially a type for the private printer rather than for the general run of publications.

The matter stands this way. I have recut the matrices — in what are called "masters" from which the Monotype Co. can make electro *display matrices and which Mr. Harris could cast on his monotype or Thompson casters. These electro mats will cost me practically $100 — possibly a trifle more depending on the actual number of characters — and I would want at least $600 for the exclusive rights to the design or $700 for the matrices & rights. $250 down and the balance in 1, 2, or 3 monthly payments from the time when mats were ready for delivery (if time is of any object to you) I might accept 50 less for all cash. This price is practically 1/2 my usual charge for a type and if not disposed of now will go into my specimen of new faces & sold through Continental Typefounders to printers, specimen now preparing.*

I am sending you some proofs of it — it has never been shown except in two little Christmas cards of my own — that is, none [type] has been sold to any printer.

Mr. Harris tells me that you met my friend Bruce Rogers in Eng — you probably met also my friends Geo. W. Jones & Stanley Morison. Rogers is here now and is returning to Eng this month to finish the Bible now printing at Oxford. We — Mrs. Goudy & I — have just completed the composition for him in Deepdene Italic (16 pt) for a privately printed edition in limited number of the letters to 'BR' from T E Shaw (Lawrence of Arabia) on the making of the "Odyssey."

From what Mr. Harris tells me regarding you it would be pleasant to know that the type, which in a sense is an unique design, had an appreciative owner & a good home. There are a few slight changes in two or three characters I would make and possibly add two or three to add to its value & usefulness. Additional sizes could be added as needed & a reasonable charge made for cutting.

I shall be glad to hear from you. Yours very truly,

Fred W Goudy

There was no possible way I could raise $700 at that time, and I was devastated at the lost opportunity of having a private press type of my own. Ed Grabhorn ultimately bought the typeface for himself. He renamed it "Franciscan" and printed many handsome books using it. After Bertha Goudy's death in the fall of 1935, Fred made frequent trips to California, having an old girl friend, Florence Duval, to whom he was quite attached. She lived in the San Gabriel Valley, not far from Claremont, where Scripps College was located. I saw quite a bit of him and often took him out to Scripps College. It was a girls' school, and he enjoyed the attention they paid him.

At about that time he wondered "why the head of some great university has not tried to gain greater distinction for its publications by acquiring a type which should be the university's own." Samuel Farquhar, the head of the University of California Press, had the same idea and on December 30, 1936, wrote Goudy, "One of the Regents of the University who is much interested in printing and the development of the Press suggested to President Sproul that we should have our own type face. Would you consider designing a face for us?" He was agreeable. The University wished it to be completed by 1940 to be used in an appropriate memorial for the 500th anniversary of the birth of printing. This gave him an additional reason for visiting California.

The traveling exhibit of the Fifty Books of the Year selected by the American Institute of Graphic Arts came to Scripps College. There was an opening reception and since two of my books were included in the exhibit, I was invited. The show of books made quite an impression on the girls, and I was asked if I could teach them something about printing. I agreed and gave a six-week course at my shop for a group made up of students from Scripps, Occidental, Mills, and Stephens colleges. Goudy spent a good deal of time in California during the winter of 1939–40, first at Berkeley where the press was printing its first book in the new type, a book written by him, and then in southern California. Among many receptions for him was one at Scripps. The Class of 1941 was already planning as its class gift to endow a press. They had accumulated $150. As Mr. Goudy was telling about designing the California type, one of the girls whispered in the librarian's ear asking if she thought Mr. Goudy might design a type for Scripps. She added that quite possibly her grandmother might pay for it. The librarian, Miss Drake, suggested she should first call her grandmother to find out if she were willing and then ask Mr. Goudy. Grandmother was willing to give $1,000, and the intrepid young girl immediately asked Mr. Goudy if he would design a Scripps College type, which the Class of 1941 could give to the college. He said he would consider it. Later on, after he had agreed, he laughed as he told me, "Those girls really ensnared me. I charged the University of California $10,000 for their type and now I'm doing a more difficult design for only $1,000." Someone asked him how he designed a type. It was a question he'd probably been asked many times before so he had a quick answer, "I just think of a letter and then draw a line around it."

We had seen preliminary proofs of the new type and

had made a few minor suggestions, but it was not ready for graduation time in May 1941 when elaborate preparations had been made for the class presentation. Regardless, Goudy came West for the occasion. Since the press would be handicapped operating with only his type, I gave the college my Washington handpress and arranged for some other essential material. Goudy finished his work on the type during the summer and sent the matrices to McKenzie & Harris in San Francisco to cast up a trial font, which they sent down to me. Another dinner was planned for September 18, 1941, and Goudy came West again. We had planned a keepsake for the occasion, and Goudy came to my plant where we set it up and then took it out to the college and locked the type up on the handpress. During the course of the evening, Goudy was the first to print a keepsake from his new type, followed by all of the other guests. I spent most of my Saturdays during the next few years teaching a class in printing at the college, showing a group of dedicated girls how to set type and use the Washington handpress.

Janet and I were at a house party at the Stephens Ranch in the foothills below Mt. Baldy on the weekend of December 6, 1941. All was fun and gaiety with plenty of Ramos Fizzes and Bloody Marys until one of the guests turned on the radio to get a bit of music. Instead all we heard was the terrible news about the bombing of Pearl Harbor. Our immediate reaction was one of disbelief and then of horror and fright. We envisioned an immediate invasion of California. The party broke up as we all headed for our own homes to prepare for the worst.

The effect of the war on our printing business was calamitous. Institutions and businesses practically stopped buying printing. At the time our staff was small, consisting only of Gregg Anderson; Joseph Simon, our compositor; Earl Myers, the pressman; and me. Someone had to go, and since I had a family to feed and had been offered a job at Douglas Aircraft Company, it was decided that I should be the one. I was given a position in the Santa Monica plant of producing training manuals to help the great number of untrained women and elderly men who were being brought in to replace the younger men who were going off to war. It still amazes me how quickly these previously untrained people adapted to their new responsibilities and could be trained to do their part in making planes. We had to start with basics, teaching through illustrated booklets how to drill, rivet, use a hacksaw, or read a rule. We had a splendid team of writers and artists doing the text and illustrations, including a former Rhodes scholar, several college professors, and some well-known artists. I made the layouts and handled the production.

At that time I was living in La Cañada, having moved there in 1941 soon after our second son, Duncan, was born. It was a lovely spot, a couple of acres covered with large oak trees bordered by a stream. The stream was a dry playground for the kids most of the year, turning into a roaring river during the rainy months. I really couldn't have afforded to buy it for the $7,500 asked, but the owner was persistent and took a mortgage on my old farmhouse in Los Angeles to make sure we bought it.

When I accepted the job at Douglas in Santa Monica, I hardly realized that I'd have a four-hour commute every day. It meant getting up each morning at four o'clock, with Janet preparing my lunch, and not getting home for dinner until around eight in the evening. We worked six days a week and I hardly saw the sun except on Sundays. In those days of gas rationing, I would stop on the way to pick up fellow workers and arrive at work with a full load. It was an interesting job, even though wearisome. The acres and acres of the many buildings of the plant and the parking lots were covered with a canopy of cloth on which was painted as camouflage neighborly streets and homes. Occasionally I'd get over to the Long Beach plant, which was newer and had some odd innovations. One in particular I shall always remember. The lighting in the shops made everyone look hideously cadaverous, which diminished wasteful sociality between the sexes.

The hours were exhausting, and when I was offered the position as head of the production department at Foote, Cone & Belding, the largest advertising agency in southern California, I was only too happy to accept. It was a challenging and varied experience in which I learned to work under the pressure of inflexible deadlines and numerous crises. These I seemed able to handle, being somewhat impassive by nature. It was a fairly large operation, but I organized it with each of my assistants responsible for his or her accounts so I was free to deal with clients and suppliers unless there was trouble.

The agency at that time was a close-knit organization, a harmonious family that worked well together and played well together. I learned a lot as a member of the planning board and through association with some exceptional artists. From these artists I learned to design on a drawing board, rather than with type and varying proofs as I had previously been doing at the printing shop. It could be faster and much more versatile. These layouts also could be given to others to process. Yet my previous experience with type was invaluable as I could visualize its eventual appearance when printed.

Foote, Cone & Belding was the stepchild of a much older agency, Lord & Thomas. It had been one of the oldest American agencies, owned by Albert Lasker. He had incorporated several large companies that he owned into this conglomerate. The anticipated new tax structure Congress suggested with the advent of the war made it advisable for Lasker to remove his various companies from under the umbrella of Lord & Thomas. This he did, and being no longer especially interested in the advertising business, he divided the business among the five resident managers. The British and Canadian managers formed their own separate companies, while the three Americans, Emerson Foote in New York, Fairfax Cone in Chicago, and Don Belding on the West Coast, merged their bounty into the new firm of Foote, Cone & Belding.

I spent a pleasant eight years with the firm. The office in the old Edison Building, kitty-corner across from the Biltmore Hotel and opposite the public library, was handy to the bookstores and to the since demolished University Club. Most of the advertising fraternity belong there, and the Zamorano Club also had its clubrooms and library there. Most of us gathered at the club for lunch and a short respite before returning to work. The agency bought an astonishing amount of typesetting for its advertisements, and Don Belding suggested that we form our own company to handle it. He and I, with Bert Oliver, the office manager, soon became partners in William Caxton, Inc., which I named after England's first printer. It thrived, but eventually the agency bought us out. In 1945 I gathered together a handful of production men from the other agencies in town and we formed the Advertising Agency Production Club, which still functions after all of these years.

For some vain reason I tried to conceal my baldness by combing a few strands of my remaining hair across my brow. One fateful afternoon after a three- or four-martini luncheon with a client, the girls in my department caught me in an unguarded moment and clipped that offending strand of hair. A few days later I came into my office to find a blown-up framed picture of myself with the top of my head painted bald and the missing hunk of my hair neatly turned into a mustache. When I retired from the firm in 1950, there was quite a celebration, beginning with a black announcement. On the front cover it read, "In Memoriam." Inside it read, "Ye Printer to ye Devil has Gone. In loving memory of our beloved Ward Ritchie, who is departing the halls of FC & B, a stag wake will be held in the studios of Sass-Wilson the night of Friday, April the fourteenth, in the year of our Lord, nineteen hundred and fifty. You are invited to help paste the dummy. A charge of five dollars each will be made to cover

PU234567

the costs of booze, food and an extra case of booze. Any extra cash will be assigned to the Ritchie Hairpiece Fund. Enclose find blotter for tear-shedding purposes."

It was quite a party. One of Bing Crosby's hairpieces had been borrowed to crown and conceal my baldness. There was also an amusing gallery of pictures. An eight-by-ten picture of me had been distributed to an array of artists who had painted on the photo their concept of a perfect hairpiece for me. Their concepts varied from making me look like Robert Taylor to a Zulu tribesman. I failed to follow any of their suggestions after discovering the pleasure of no longer having to groom my locks.

During the years I was away from the press, I was still able to steer work to it and also to design a few of its books, but the operation was left primarily in what proved to be the capable hands of Joseph Simon. Gregg Anderson had been drafted, never to return, as he was killed in the invasion of Normandy. His widow, Caroline, took Gregg's place at the press and continued there for several years until she married Dr. French Fogle of the Huntington Library staff, later a professor at Claremont Graduate School. Earl Myers, who had been our chief pressman, had graduated into a successful salesman for the press. That team had not only carried the press through the war years, it developed and expanded it into a primarily commercial plant. In 1950 Earl Myers was wooed away by a competing organization. Customers' loyalties are often to the salesman who calls on them rather than to the distant and sometimes nebulous plant that produces the work. It was feared that Myers might take with him many of the press's major customers. I had a hurried plea from Simon to return to the press and help keep the accounts that Myers had served. That is how I came to leave the advertising agency and return to the printing business.

It was a far different organization than it had been when I left it eight years before. Simon was a compulsive buyer of equipment. I had always been reluctant to buy anything until I was sure there was money to pay for it, but I couldn't begrudge Simon's success and felt obliged to go along, though at times I was nervous. We incorporated under the name Anderson, Ritchie & Simon: The Ward Ritchie Press, using the AR&S name for commercial work and the Ward Ritchie name for what publishing we did. There were no longer the intimate gatherings that had made "Ritchie's Roadhouse" such fun and had attracted so many interesting visitors. I became converted into a full-time salesman and part-time designer.

During the final year that I worked at the advertising agency, I also began doing some designing for the University of California Press. Samuel Farquhar had directed the press for almost twenty years during the thirties and forties until his death in 1949. At that time the publishing division was separated from the printing department and August Frugé was selected to direct the publishing. During Farquhar's regime he, with the assistance of Amadeo Tommasini, had handled the design of all the books issued by the press. With the expansion of publishing after the division, the printing department could no longer handle the publishing volume along with all of its university work. It became necessary for Frugé to seek outside printers and designers. I was chosen as one of the first of these designers, and I found that my experience on the drawing board at Foote, Cone & Belding was invaluable in making precise layouts that any printer could follow and produce a book as I envisioned it.

August Frugé in his *Design & Printing at the University Press* tells the story of my involvement: *Feeling some urgency to show that we were not bibliobarbarians, that we knew a book from a government document, we cast our net for the best designers within reach. As the first of these, and our surest bet, we chose Ward Ritchie of Los Angeles. In skill and reputation he was on a par with the finest private printers in California or elsewhere, and he was more businesslike than most. One could hardly ask the great loners, such as Ed Grabhorn or Saul Marks, to provide layouts suitable to the commercial printers we had to use, but Ward was at home in both worlds and could provide what we needed. Over the next few years he designed about seventy-five books for us, including such great sets as* The Sermons of John Donne *in ten volumes, and six of his books were chosen among the AIGA [American Institute of Graphic Arts] Fifty Books of the Year. One could say that we hit the jackpot on the first pull.*

Janet decided to divorce me in 1946 and find a different life. We divided our two-acre plot. She and the children continued to occupy the house, while I moved into a shack on the back acre. I became an immediate architect, plumber, and carpenter, rebuilding and expanding the structure into what I found to be a very comfortable studio and living quarters. My crowning achievement was a huge fireplace, the logistics of which baffled most who saw it, even my architect friends. I had written for a government bulletin on the making of fireplaces, and though I didn't follow all of their directions verbatim, I was able to figure out the ratio of openings to chimney height, etc., to achieve what I envisioned — a large copper hood against a flat, whitewashed brick wall. It was magnificent, with the copper hood somewhat over six feet wide. I made the place airy and bright with many paned windows. I found a wonderful door from an old Prohibition days speakeasy and bought Grant Dahlstrom's handpress, which decorated the room. Outside I laid flagstones under the trees and eventually put in a swimming pool. It was a great party place. On one such occasion some of my waggish friends, unbeknownst, erected two poles at the street entrance to my long driveway and hung a sign, "Ward Ritchie Memorial Park," and then planted white crosses all along the winding driveway. Local residents who had passed by are still wondering whatever happened to the "park."

For four years I was a happy, rollicking bachelor, and then I met Marka. In the meantime Janet had moved away. I was asked if I would accompany a young woman to the Gardenia Ball of the Assistance League at the Huntington Hotel in Pasadena. After I had accepted, my friend assured me that I would find her most agreeable as she was a Phi Beta Kappa and an ex-school teacher. I was apprehensive when I went to pick her up at her home in San Marino. I shouldn't have been. I found her to be a lovely young creature, a brunette with soft brown eyes. I soon discovered that I was more than casually attracted and was writing love poems to her.

This song that I must now sing to you
will tell you of many things –
of the mountains, holding out the north,
that come their ragged way
down through the panelled windows here –
of the sky that's also there
and yellow butterflies
dancing on its rim –
of trees, whose emboldened fruit
glisten red through their sheltering green –
and of the soft-skin mole
whose drunken path
rumples the dew-sparkled grass.
It is June now for us,
with all of June's gracious beauty.
And yet when you are away from me
the days hold a drabness in their sky,
for only there where you may be
can I aught of any beauty see.

We were married four months after I met her on that blind date, and she brought her three young boys, Mark, Jan, and Peter, with her to La Cañada. My oldest son, Jonathan, joined us. Duncan continued to live with his mother, who had also remarried. We built on a kitchen and a dormitory for the boys and thus began a marriage of thirty-one years until Marka died of lung cancer in 1981. Marka made friends with my friends very soon. She was an especially good cook and enjoyed cooking. She had this in common with our good friends Philip and Helen

Evans Brown. Philip was a bookseller and a collector of cookbooks. Helen had been a caterer and was working at the gourmet grocery store, Balzer's, on Larchmont Boulevard in Los Angeles. They had joined me, along with several other friends, almost every Sunday for games and barbecue during the war years. It was Helen who brought me the manuscript of Robert Balzer's book on wine that originally got the Ward Ritchie Press interested in the cookbook field. Bob Balzer was the son of the owner of the grocery store and was in charge of the wine and spirits department. He made periodic trips to the wine country and had written a book called *California's Best Wines.* It was a good book, but it was a new field for us and I was undecided as to the advisability of publishing it until Helen informed me that Balzer's would place an initial order for 500 copies. We published the book, which contained, in addition to the text on wines, numerous recipes. These seemed to appeal to our customers as much as the material on wines. It sold so well that I was willing to experiment further in this new field and asked Helen Brown if she would be able and willing to write a cookbook for us. She was definitely able and certainly willing. Her first book for us was the *Chafing Dish Book,* which went into more than a half-dozen printings. It was followed by the *Patio Cook Book* and *A Book of Appetizers.* She was good and we could hardly expect to keep her exclusively for ourselves, and in no time she was besieged by eastern publishers soliciting manuscripts. They published a half-dozen of her books before her death in 1965. One of these, *The Boys' Cook Book,* she dedicated to our sons. Another,

Helen Brown's West Coast Cook Book, is in my opinion one of the best of American cookbooks. It was published by Little, Brown and Co. I was pleased when asked to design it.

As publishers we were becoming more and more oriented toward cookbooks, doing several new titles each year. Marka was also becoming involved. We became very attracted to Elena Zelayeta as a person and as a cook. Elena was blind, but still a fabulous cook, running her own little factory in San Francisco, which prepared Mexican gourmet delicacies. She lectured and gave cooking demonstrations of Mexican recipes to womens' clubs all over the country. She took copies of a small booklet of recipes with her on these demonstrations and had sold over a hundred thousand copies. One evening when she was visiting with us and the Browns, we suggested that she was worthy of a better book and she happily agreed but explained how difficult it would be for her to prepare it. Arrangements were made for Helen and Marka to drive to San Francisco and stay with her for a few weeks while she would cook for them her favorite recipes and they would take notes. It was quite an experience for both Marka and Helen, not only for Elena's hospitality but for the opportunity to watch how this remarkable woman could cook her unusual dishes being completely blind. For two weeks Elena cooked constantly while Marka took notes. One problem was in determining quantities of the different ingredients. Elena had an instinct of how much of each to use, as she let the flour, the milk, or the spice flow through her fingers. She could feel what was right, but whether it was half a cup or three-quarters of a cup she could not see to tell. Marka and Helen had to carefully measure everything.

Marka returned home with a bulging notebook and settled down to write the book. She gave it an unusual twist. She described each particular dish as if she were Elena explaining about the procedure and ingredients. Marka would then add her own observations and commentary. It was a fun book to design and was chosen as one of the AIGA Fifty Books of the Year as had been Balzer's *California's Best Wines* and Helen Brown's *Chafing Dish Book.* Marka was also involved in several other cookbooks we published. Osborne Putnam Stearns submitted two manuscripts for us to consider, one a travel book, the other a cookbook. Stearns died in the meantime and Marka took over, revising and combining the two books into one, which we published as *Italy on a Platter.* Marka's great love was Mexican cooking. She had spent some of her undergraduate college days at the University of Mexico, and she retained a love for everything Mexican. The manuscript of a Mexican cookbook was submitted to us. We were about to reject it when Marka asked if she might work on it. She did and we published it as *The Food and Drink of Mexico.* She was never completely happy with it.

Of all of the forty or fifty different cookbooks we published, none has a better story than *The Cookout Book.* Among the many products Henry Kaiser produced was Kaiser aluminum foil. To promote it, he or his advertising department conceived of having a national cookout contest for men only, in which they were to use Kaiser foil. The first one was in 1959. It was nationally publicized, and recipes came in from all over. From these, twenty-five were selected and the lucky contestants were flown to Hawaii to Kaiser's hotel there to cook and compete for a prize of $10,000. After ten years it was proposed to have a cookbook containing a selection of the recipes that had been submitted. One of my former associates at Foote, Cone & Belding, Mike Corcoran, was now with the advertising department at Kaiser's headquarters in Oakland. He called me to come up there in January to see if I could produce a book for an extensive campaign they were planning for May, just before the barbecue season began. It was a challenge that I accepted, not realizing then what it would entail. We were to make a selection of recipes from those that had been submitted over the past ten years and produce a book.

We had moved by that time from our place in La Cañada to a much larger house in nearby Flintridge. It had a large patio area in the rear of the house, which proved to be most fortunate. I hadn't realized how popular this contest had been over the years until a truck arrived full of crates of recipes. I had also requested fifteen or sixteen rolls of aluminum foil for us to use in testing the recipes. Instead there arrived fifteen large cartons of foil. I thought Marka might leave me then and there as she watched all of this being unloaded. I had already made preliminary layouts, selecting a square format eight and one-half by eight and one-half. I had asked Philip and Helen Brown to write a long introduction on "Barbecue Cooking and Entertaining." Also I had alerted George de Gennaro that I would need eight or ten color shots of the dishes, and Harry Diamond that I wished him to do fifty or more spot drawings to enliven the pages. But the main job was to sort and select from the thousands of recipes we found in the crates. This task rested with Marka and me, but mostly with Marka.

The recipes came in all shapes and sizes, from elegantly bound books and illuminated manuscripts to scrawls written on scraps of paper. We went through them one by one. We were able to eliminate the majority in the reading, and finally to sift the mass down to two or three hundred, which we felt deserved testing. We parceled some out to friends to test for us, and then set up four barbecues in the patio for our own testing. Each morning we would schedule the day's program. We would shop for the ingredients we would need, light the charcoal, and start cooking and testing. Most of the work fell upon Marka as I still had other commitments at the shop. We finally selected about a hundred recipes we felt were deserving. It was an interesting medley. Marka insisted on including a few oddballs to enliven the book, such as "Barbecued Beef Head, Old-California Style," along with "Lamb Chops for Grandma Who Likes Martinis." It all jelled beautifully, the text, the illustrations, and the design. We managed to meet our May deadline. Mr. Kaiser was happy and the AIGA chose it as one of the Fifty Books of the Year. It was later included in a show in Europe as an example of American bookmaking.

I had always enjoyed the work I was doing. My philosophy has been that we only have one life. Try and make it useful if you can. Enjoy each day as long as you have any days. The time came when I was no longer enjoying what I was doing. I began to recall the joy of the early days of my printing — the days of my apprenticeship in Paris, the nights and days I spent in the little shop in back of my family home in South Pasadena, and of the camaraderie of "Ritchie's Roadhouse." I decided the time had come to try and recapture some of the exaltation of those days. So in 1972 I retired from the press I had started just forty years earlier. I wanted to again play with printing, to experiment and to write.

Remembering the pleasure I'd had puttering with a handpress in those early days, I began to hunt for one. We had been planning a trip to England and I inquired of companies there. I had some rather good prospects, but not long before I arrived in London a couple of rare handpresses were auctioned at Sotheby's at ridiculous prices, and every dealer in London had jumped the prices I had been quoted. This annoyed me. I returned home without a handpress, a bit discouraged. However, in the accumulated mail was a packet of printed pieces from the Amalgamated Printers Association, among which was one offering an Albion handpress for sale to the highest bidder. It was in Boston. I phoned, asking if the press had been sold. The answer was "no," but there had been some offers. By then I was frantic for a press and said, "I'll give you a hundred dollars over your best offer." The immediate reply was, "sold." He asked for my name and when I replied Ward Ritchie, he laughed. "Ward," he said, "this is Jonathan Macy." I had known Jonathan and his parents, Helen and George Macy, who ran The Limited Editions

Photo by Amanda Blanco

Ward and his press

Club. Jonathan had even had a small reception for me at the Harvard Club on one of my visits East. I paid him $600 for the press sight unseen. It cost me more than that to have it packed and shipped to California. It arrived, to my delight, in good working condition.

Jonathan told me the story of how he had come into possession of the press. After his mother and father had died and the Limited Editions Club had been sold, he decided to start a new club, with each book being printed on a handpress in the tradition of William Morris. John Dreyfus in England had selected a press for him and shipped it to Boston. It didn't take Jonathan too long after starting to print a 300-page book by hand to realize that his plan was impractical, and that is why he was disposing of the press.

Over a period of years we had printed several books for the Limited Editions Club. The first one caused some unhappy moments. When George Macy picked Edward Bellamy's *Looking Backward* for his 1941 list, he envisioned a book that might have been printed in the year 2000 when the book was supposed to have been written. He chose Merle Armitage to design it. Merle was a maverick in the book world who was ignoring all tradition in his designs. Nowadays this book would not appear outrageous, but at that time the book was anathema to the traditionalists who made up most of the membership of the club. Armitage and Macy exchanged some harsh words, with me in the middle getting a flood of letters from each of them. Somehow I managed to remain neutral and friendly with both. Armitage continued to have me print several of the books he designed, and Macy elected to make me a contributing editor of his publication, *The Dolphin: A Periodical for All People Who Find Pleasure in Fine Books.* But it wasn't for another several years that he asked me to do another book for the Limited Editions Club. He and Helen often came to spend some of the winter months at La Quinta in the Palm Springs desert and in Beverly Hills. Marka and I would see them on these visits, and on one occasion we introduced them to Millard Sheets, a southern California artist. He charmed the Macys immediately. Both versatile and talented, he painted, he designed buildings, he ran the art department at Scripps College, and he was director of the Otis Art Institute. One of his best-known achievements was the mammoth mosaic on the facade of the library at the University of Notre Dame. George Macy almost immediately arranged for us to collaborate on producing Robert Louis Stevenson's *The Beach of Falesá.* Millard's great problem was that he would get enthusiastic about a project, but when something else came up, he would sidetrack it. This happened so often during the production of this book that it took almost a year to produce it. It was better when we did Joseph Conrad's *The Nigger of the Narcissus* because he had only to do full-page paintings rather than making sketches to fit my layouts.

In 1956 when Marka and I visited the Macys at La Quinta, George was bothered with stomach pains. It was soon determined that he had cancer. He was operated on in Los Angeles and was recuperating at the Beverly Wilshire Hotel when Paul Landacre and I collaborated on a get-well card for him, using one of Paul's wood engravings. When Marka and I visited him a few days later, he was bothered by a situation he found himself in with Lucretius' *De Rerum Natura.* He had earlier commissioned Landacre to illustrate the book and Bruce Rogers to design it. Rogers seldom liked others' illustrations for his books and he was especially unsympathetic with the abstract engravings Landacre had made. He had submitted his layout to George with some of his own decorations. George was in a quandary and asked me if I would be willing to design and print the book for him. This was embarrassing to me, but George assured me that he would pay Rogers his usual $2,000 fee and give him another more compatible book to do. I agreed. I was able to show him proof pages with Landacre's blocks before he left for the East, but he died before the book was finished. He left with me his work table and file. He always brought his work with him on his vacations. There was a little note attached, written on one of the labels Bill Dwiggins had designed for him, reading:

> Bless the house this file is in:
> Keep Old Pappy off the gin;
> Help Mamma to put on weight;
> Damn the boys when they are late.
>
> G.M. 4/30/56

After George's death we did several more books for the Limited Editions Club before it was sold. These included *Call of the Wild, American Indian Legends,* and a group of Conrad stories called *Youth, Typhoon and End of the Tether.*

Jonathan Macy's press finally arrived in Los Angeles. It

was in good shape for having been built in 1835, but there is little to go wrong with an Albion handpress. At that time we had a home in Pasadena and a weekend home in Laguna Beach. The smog that seemed to be getting worse in Pasadena was affecting Marka's bronchitis. I had to make a decision as to where to install the press and came to the conclusion to sell the Pasadena place and move to the beach. The beach house was on a slope not far from the ocean. There was a fair-sized recreation room on the lower level, which I thought would make a pleasant studio and pressroom. I hadn't envisioned the problem of getting the heavy press down the slope and around the narrow path and into the low-ceilinged studio. When it was finally installed, the weary movers informed me that if I ever wished to move the press again, I'd first have to tear the house down.

That was in 1974. This was a much smaller house than we had in Pasadena, and it became necessary for me to part with three or four thousand books, keeping only my press books and a couple of thousand books about books, which I needed as a working library. I could hardly wait after the press was installed to start experimenting with nature printing, *i.e.*, printing images directly from ferns and leaves. I was planning to use them as decorations for a book of MacIntyre's unpublished poems, which he had sent to me over the years. However, I put it aside when Saul Marks died. He had been the finest traditional printer in southern California. John Dreyfus had written a fine memorial about him, and Jake Zeitlin asked me if I would print it. It had been nearly forty years since I had actually done any manual printing or set type. I was rather rusty at it and had to relearn much that I had forgotten. I managed somehow to fumble through and complete the first booklet from the Laguna Verde Imprenta. This was the name I had chosen for my new press — Laguna for the town; Verde (which is Spanish for green) for Emerald Bay where we lived; and the Spanish Imprenta for printing shop. The English wood engraver Leo Wyatt had cut a printer's mark for me, which I used here for the first time, and Zeitlin had given me an original Eric Gill engraving, which I used as decoration on the title page. I jumbled the type in my ineptitude and pushed it back into place incorrectly, so many of the copies have a bungle that probably caused the meticulous Saul Marks to turn several times in his grave.

Many of our friends and neighbors were curious about the hulking machine we had moved in, so we had an "iron monster party" and gave them all a chance to try their hand printing on it. I concocted an axiom and set it in type, and each was allowed to print it as a souvenir: "This Body I wear is much too fragile for the Life I lead." The first few pieces I printed on the press were of minor importance. I was learning the whole process again after so many years. I had no intention of doing press books and reprinting the classics in the tradition of William Morris and Cobden-Sanderson. I was more interested in doing original and unpublished material, mostly my own derived from lectures I had given. These included books on Robinson Jeffers and the Ashendene Press, *Adventures with Authors* and *The Poet and the Printers,* together with

Laguna Verde
IMPRENTA
1975
1987

A Late Offering of Quince
by Peter Lum Quince
1990

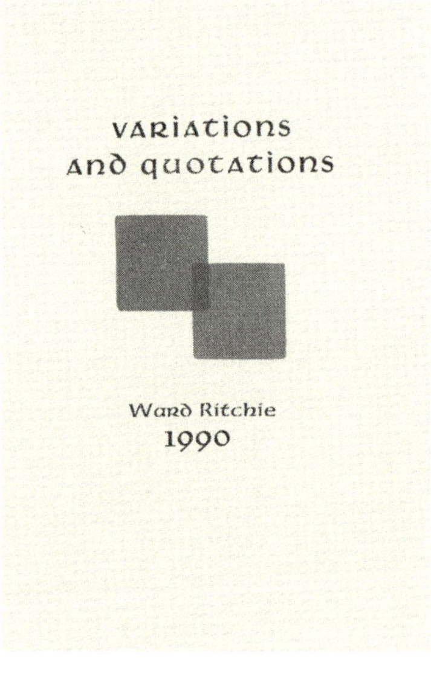
VARIATIONS
AND QUOTATIONS
Ward Ritchie
1990

157

Laguna Verde Imprenta

books about Merle Armitage, Jane Grabhorn, and Lawrence Clark Powell. Glen Dawson later gathered many of these together and published them in a book entitled *Of Bookmen & Printers: A Gathering of Memories.*

Between the years 1975, when I began printing on the handpress, and 1987 I had done twenty-six items of some minor importance. I was eighty-two years old with failing eyesight and concluded I should do a bibliography commenting on each of the items and then call it quits. I printed a rather elaborate book, but I didn't call it quits. I found I wasn't quite ready to abandon my printing. Glen Dawson had previously suggested that I should do a miniature book using some of the small engravings Landacre had made for me. I mixed them with a few poems by Quince and printed fifty copies, calling the book *Landacre and Quince.* Joseph D'Ambrosio bound it and Dawson bought the whole edition and sold all copies in one day. Over the years I had jotted down quotations that appealed to me. In England I had seen a book with abstract illustrations created from simple blocks of engraver's wood. On returning home I started playing around with rules, blocks, and a variety of colored inks, finally combining them with a dozen and a half of these words of wisdom to make a commonplace book called *Variations & Quotations.* D'Ambrosio created an original and harmonious box binding for the book. Finally, before I became diverted into the writing of this account, I printed a new batch of poems by Peter Lum Quince called *A Late Offering of Quince.*

A couple of years after Marka's death, Powell and I

gathered a sheaf of our old friend MacIntyre's early lyrics to print as a keepsake for the members of the Zamorano and Roxburghe clubs. At about the same time there appeared a glowing pictorial story of the former motion picture star Gloria Stuart in the magazine section of the *Los Angeles Times.* It was a pleasant surprise. I hadn't seen much of her for years, though we had once been such close friends. I was pleased to learn that she was still active and as beautiful as ever. The story told of her achievements as an artist after she had forsaken the screen at her husband's request. Her paintings had been exhibited in many galleries, including the Hammer Gallery in New York.

Since she had been a friend of Carlyle MacIntyre in our early years together, I mailed her a copy of his *October Songs,* which we just had printed. She replied with a pleasant letter suggesting if I should ever be in Los Angeles she hoped I would drop by for old time's sake. Some time later I did. It was a dramatic reunion after so many years apart. She wrote of that moment this poem:

And so it came to be,
himself walked up the path,
spring-shaded with my elm's new leaves.
I watched behind the heavy silken knots
masking my windowed door,
and opened it before he climbed the steps,
making him welcome
as the bronzy bells tinkled on their long fringed knots
and we smiled uncertainly at each other,
and then so lightly clasped each other's hand.

160

He walked into my parlor,
bright with lamps and flowers,
and then into the bar
warmed with a burning open fire,
a mirrored crystal alcove
full of all the alcoholic props.
He offered me a book,
a gift of his own making.
My cats then stalked around his feet,
antennaed tails held high, and soon departed.
He mixed his vodka on the rocks, with a lemon twist.
I poured my wine
and we began to talk, to laugh,
to reminisce, with very few regrets,
about our friends and lovers
(and with some surprises!),
our own two sorrows, past now,
not heard, or known about
by either thee or me —
nor shared at any time.
I know I loved you instantly.
I know I felt your warmth
and depth, your valor
and your love
so very quickly,
there was no need for games.
There never has been.
That is what I treasure most.
My heart an open book
upon your open hand.

When she saw my handpress and the books I had printed on it, she said, "I can do that." And she did. She bought a press of her own and quantities of type. She produced books under the imprint of Imprenta Glorias. They are innovative far beyond my ken. She has become a stimulating companion for these fading years.

As I look back in memory upon my life in printing, I find it separated into four distinct parts — my Four Ages of Printing. Originally were the learning years in Los Angeles and Paris, followed by the Ward Ritchie Press days of the 1930s. The third stage would be the Anderson, Ritchie & Simon period during the fifties and sixties, and finally the handpress days that I now know.

The years of life have not all been joyous, but I have been happy. And as Peter Lum Quince concludes:

The world began
when I was born
and it will end
when I am gone.
The sun, the sky,
the night, the day
will cease to be
along with me.